The 1946 and 1953 Yale University Excavations in Trinidad

YALE UNIVERSITY PUBLICATIONS IN ANTHROPOLOGY

The Yale University Publications in Anthropology series, published by the Yale University Department of Anthropology and the Peabody Museum of Natural History at Yale University, is supported by the Theodore and Ruth Wilmanns Lidz Endowment Fund for Excellence in Scholarly Publications, dedicated to the dissemination of scholarly research and study of the world and its cultures.

The Yale University Publications in Anthropology series embodies the results of researches in the general field of anthropology directly conducted or sponsored by the Yale University Department of Anthropology and the Yale Peabody Museum of Natural History Division of Anthropology. Occasionally other manuscripts of outstanding quality that deal with subjects of special interest to the faculty of the Department of Anthropology may also be included.

For a complete list of available titles in this series visit www.yalebooks.com.

The 1946 and 1953 Yale University Excavations in Trinidad

Arie Boomert
Birgit Faber-Morse
Irving Rouse

With contributions by
A. J. Daan Isendoorn
Annette Silver

Number 92

Published by
the Yale University Department of Anthropology
and the Yale Peabody Museum of Natural History

Distributed by
Yale University Press
NEW HAVEN AND LONDON

Yale

YALE UNIVERSITY PUBLICATIONS IN ANTHROPOLOGY
Number 92

Rosemary Volpe, *Publications Manager*

Front cover: Excavation of Trench 2 at the St. Joseph 2 site, Trinidad, in progress in 1953. Photograph by Irving Rouse. Courtesy of the Division of Anthropology, Peabody Museum of Natural History, Yale University.

Back cover: Irving Rouse (at far right), Fred Olsen (at left) and José M. Cruxent (center) visited the dig at Cedros, Trinidad, in September 1969 to collect samples for radiocarbon dating. Courtesy of the Division of Anthropology, Peabody Museum of Natural History, Yale University.

Title page: A Palo Seco complex modeled pottery fragment from Excavation 2 of the Quinam site, Trinidad (YPM ANT 184208; see Figure 55D). Illustration by Armand Morgan.

For submission guidelines visit peabody.yale.edu or send inquiries to:
Publications Office, Peabody Museum of Natural History, Yale University,
P. O. Box 208118, New Haven CT 06520-8118 USA; peabody.publications@yale.edu

ISBN 978-0-913516-28-7
ISSN 1535-7082
Printed in the U.S.A.

Library of Congress Cataloging-in-Publication Data

Boomert, Arie, 1946-
  The 1946 and 1953 Yale University excavations in Trinidad / Arie Boomert, Birgit Faber-Morse, Irving Rouse ; with contributions by A.J. Daan Isendoorn, Annette Silver.
      pages cm. -- (Yale University publications in anthropology ; Number 92)
  Includes bibliographical references and index.
  Summary: "This first detailed review of indigenous cultural development on the island of Trinidad, from pre-Columbian through historical Amerindian sequence, reports on previously unpublished archaeological excavations there in 1946 and 1953 by Yale's Irving Rouse, and analyses this work in light of ceramic evidence, subsequent research and the wider Caribbean context"--Provided by publisher.
  ISBN 978-0-913516-28-7 (pbk. : alk. paper)
1. Indians of the West Indies--Trinidad and Tobago--Antiquities. 2. Excavations (Archaeology)-- Trinidad and Tobago 3. Trinidad and Tobago--Antiquities. 4. Ethnohistory--Trinidad and Tobago. 5. Rouse, Irving, 1913-2006. I. Faber-Morse, Birgit, 1935- II. Rouse, Irving, 1913-2006. III. Title.
  F2119.B658 2013
  972.983'01--dc23
                                                                        2013009162

∞ This paper meets the requirements of ANSI/NISO Z39.48-1992 (Permanence of Paper).

10 9 8 7 6 5 4 3 2 1

Dedicated to Benjamin Irving Rouse (1913–2006),
"the father of Caribbean archaeology"

Contents

This report discusses the results of archaeological excavations undertaken by Irving Rouse at a series of prehistoric to protohistoric sites in Trinidad in July and August of 1946 and, assisted by John M. Goggin and Goggin's wife Rita, in August and September of 1953. Rouse's 1946 research took place under the joint auspices of the Yale University's Peabody Museum of Natural History and the (now defunct) Archaeological Section of the Historical Society of Trinidad and Tobago, at the time led by John A. Bullbrook. The 1953 investigations represented a joint project of the Yale Peabody Museum, the Historical Society, and the Graduate School of the University of Florida in Gainesville. Rouse's investigations formed part of Yale's Caribbean Anthropological Program, which was initiated in the 1930s by the Venezuelan work of Cornelius C. Osgood and George D. Howard next to the research by Froelich G. Rainey and Rouse in the Greater Antilles. The wish to connect the prehistoric cultural sequences devised as a result of Yale's archaeological research projects in the West Indies and on the mainland was the impetus for Rouse's work in Trinidad.

Thus far only preliminary accounts of his Trinidad excavations have been published (Rouse 1947, 1953) and, clearly, reporting on Rouse's research in the island is long overdue. His involvement with field research in Venezuela after the Trinidad work and subsequent similar investigations elsewhere, as well as teaching obligations, prevented Rouse from finishing his analysis of the material remains he had encountered at the Trinidad sites. The present report, based on our joint study of Rouse's finds more than sixty years after recovery, finally fills this gap.

Unfortunately, it was not given to Rouse to see this study finalized. His sad passing in February 2006, though expected, still came as a blow to his co-authors, just as it did to our colleagues in the West Indies and beyond. To all of us he will be remembered as by far the most important scholar, indeed the father, of Caribbean archaeology and ethnohistory.

Rouse's 1946 project in Trinidad was sponsored officially by Yale University, the Viking Fund of New York, and the colonial government of Trinidad and Tobago, the latter largely because of the personal interest in its progress of Sir Bede Clifford, then governor of the two islands. Administrative arrangements necessary for implementing the project were taken care of in Trinidad by C. Y. Shephard, then president of the Historical Society of Trinidad and Tobago, W. N. Foster, who was Government Inspector of Mines, Major (then Captain) J. E. L. Carter, and especially Bullbrook, of the Historical Society's Archaeological Section. In addition, both in 1946 and 1953 Bullbrook received Rouse's equipment, prepared living quarters and laboratory space in his home, hired labor, and secured the necessary permits to excavate. Bullbrook's assistance enabled Rouse (and Goggin) to accomplish more than would otherwise have been possible. In 1946, Bullbrook's employer, the

Trinidad Petroleum Development Company, Ltd., lent equipment and allowed the use of its facilities through the good offices of its general manager, H. A. Bennett, and its chief geologist, G. W. Halse. Furthermore, other geologists connected with Trinidad's oil industry provided information concerning sites, notably Kenneth W. Barr, C. C. Crossfield, C. C. Wilson, and Louis Winkler. Acknowledgment is due to the following companies for granting Rouse permission to work at the sites excavated: Trinidad Government Railways (Bontour); the Trinidad Base Command of the United States Army (Cedros); George F. Huggins Co. (Erin); Nariva-Cocal, Ltd. (Ortoire); Trinidad Petroleum Development Co., Ltd. (Palo Seco); and Trinidad Leaseholds, Ltd. (Quinam).

Whereas Rouse was prevented from publishing his Trinidad research by teaching obligations and field work that took him to various parts of Venezuela and the Lesser Antilles, Goggin died before he could analyze the results of his 1953 investigations with Rouse in Trinidad. Subsequently, Peter O'Brien Harris, then of the Historical Society of Trinidad and Tobago (South Section), Pointe-à-Pierre, Trinidad, was the first who expressed an interest in Rouse's Trinidad collections at the Yale Peabody Museum, examining the latter's Cedros materials during a brief stay at Yale in the summer of 1973 (Harris 1974a, 1978). He was followed by Boomert, who investigated part of the Trinidad collections during a five-week visit to Yale University in July and August of 1982. At the time, Boomert was attached as a senior research fellow to the Department of History at the University of the West Indies in St. Augustine, Trinidad. He compared materials he had excavated in cooperation with Harris at various late-prehistoric sites in Trinidad with those encountered by Rouse and Goggin. Thanks to the cooperation of Charles H. Fairbanks of the University of Florida, the collections from Trinidad made by Goggin, kept in the Florida State Museum since 1953, could be transferred to the Yale Peabody Museum. During Boomert's stay at Yale, he especially studied the late-prehistoric finds made by Rouse and Goggin at the Quinam, Bontour, St. Joseph 2, and Mayo sites in 1946 and 1953. His study resulted in an analysis of the Late Ceramic Guayabitan subseries of Trinidad and the definition of the Ceramic–Historic Mayoid series (Boomert 1985). For financial support of his 1982 research at the Yale Peabody Museum, Boomert wishes to acknowledge the University of the West Indies Department of History.

Not until Rouse and Faber-Morse had published their study of the cultural remains Rouse encountered during his work in Antigua in the 1970s did both seek ways and means to publish a final report on Rouse's Trinidad excavations comparable to their report on Antigua (Rouse and Faber-Morse 1999). They contacted Boomert, who agreed to collaborate on the project. Accordingly, Rouse, Boomert, and Faber-Morse undertook renewed research of Rouse's Trinidad collections in 2004 and 2005. Subsequently, Annette Silver joined the project to report on the historic artifacts, taking previous general identifications to a more detailed level.

Thanks to a Yale Peabody Museum travel grant, a four-week stay in New Haven in April 2005 allowed Boomert to renew his acquaintance with the Yale

Trinidad collections, this time concentrating on the materials from the St. John, Ortoire, Cedros, Palo Seco, Quinam, and Erin sites. We are grateful to Frank Hole, Richard L. Burger, Roger H. Colten, Maureen P. DaRos of the Yale Peabody Museum of Natural History, and all at the Yale Peabody Museum for facilitating and assisting our study. We were also able to select specimens for technological and petrographic analysis by A. J. Daan Isendoorn at the Ceramic Laboratory of the Faculty of Archaeology of Leiden University, The Netherlands.

Boomert wishes to acknowledge the National Archaeological Committee of Trinidad and Tobago, Port-of-Spain, Trinidad, Leiden University, and the Stichting Nederlands Museum voor Anthropologie en Praehistorie, Amsterdam, The Netherlands, for providing grants enabling the Cedros investigations in summer 2005. In addition, thanks are due to Arthur Sanderson of Petrotrin, Ltd., Santa Flora, Trinidad, who mediated in the providing of lodging facilities for the team at Petrotrin's Forest Reserve plant.

We are indebted to Corinne L. Hofman, Leiden University, for arranging for the processing of the Yale Peabody Museum pottery samples and the additional samples from the University of the West Indies selected by Boomert as part of Hofman's project "Mobility and Exchange: Dynamics of material, social and ideological relations in the pre-Columbian insular Caribbean." For providing recent information on the Trinidad sites excavated by Rouse and Goggin, Boomert and Faber-Morse wish to acknowledge Peter O'Brien Harris (Maraval, Trinidad), Keith O. Laurence (University of the West Indies), the late Archibald S. Chauharjasingh (La Romain, Trinidad), David D. A. Maharaj (San Fernando, Trinidad), Basil A. Reid (University of the West Indies), and Nicholas J. Saunders (University College London, England). Thanks are due also to the National Archaeological Committee of Trinidad and Tobago, Port-of-Spain, Trinidad, and the Stichting Nederlands Museum voor Anthropologie en Praehistorie, Amsterdam, The Netherlands, for financially supporting an extended stay by Boomert in Trinidad in August 2004, which made it possible to inspect the present state of preservation of the sites discussed in this report. Boomert wishes to thank Peter O'Brien Harris for accompanying and assisting him during this trip to Trinidad, for facilitating his stay in the island, and finally for his willingness to read and comment on parts of the first draft of this report. Regarding the present report, we thank illustrator Wendolyn B. Hill (Yale University), photographer Jerry Domian (Yale University), publications manager Rosemary Volpe (Yale Peabody Museum), and illustrators Erick A. van Driel (Leiden University) and Joanne F. Porck (Leiden University) for contributions to the volume's ultimate form.

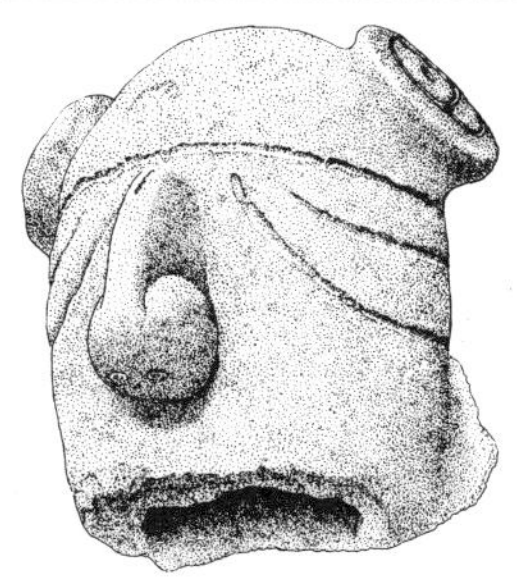

CHAPTER ONE

INTRODUCTION

Geographical Setting

Trinidad is a continental island that was separated from the South American mainland by the global rise of sea level in the post-Pleistocene era. Its once existing land bridge with the much smaller island of Tobago, which is equally situated on the South American shelf, was broken a few thousand years earlier, probably shortly after the Last Glacial Maximum. Today Trinidad is the fifth largest island in the West Indies and forms the southernmost link in the chain of islands constituting the Lesser Antillean archipelago. As such, it represents the natural gateway for human migration, exchange, and diffusion of culture from the mainland, particularly the Orinoco Valley, to the West Indies, as well as to and from the coastal zone of eastern Venezuela and the Guianas (Figure 1). Characterized by an extensive web of sea channels, rivers, lagoons, and estuaries, which formed the favorite channels of Amerindian communication and transport, this region constituted a wide-ranging prehistoric to early historic interaction sphere for peoples of different ethnic identities, levels of sociopolitical complexity, and cultural backgrounds. For instance, in aboriginal times, interaction between the Amerindians of the East Venezuelan littoral and those living across the Gulf of Paria in West Trinidad was much more intensive than that between the latter and the inhabitants of Trinidad's east and north coasts. Indeed, as Joseph (1838:2) notes, the waters of the Gulf of Paria "are so gentle that they are often traversed by a single man in a light canoe." The fluvial communication network of which Trinidad formed part had easy access to those of the Orinoco and Essequibo rivers on the mainland, which themselves offered means of penetration into the vast tropical lowlands of South America (Rouse 1983; Boomert 2000:2–3, 22, 2009).

Trinidad is more or less rectangular, with large promontories at its northwest and southwest corners: the Chaguaramas and Cedros peninsulas, the tips of which stretch to the South American mainland. These form the driest parts of the island. While north and east Trinidad receive most precipitation, the island's western portion occasionally experiences dry season droughts lasting up to eight months. In years of excessively arid conditions the incidence of widespread, mainly human-induced, bush fires is high. The other extreme is reached in the wet season when

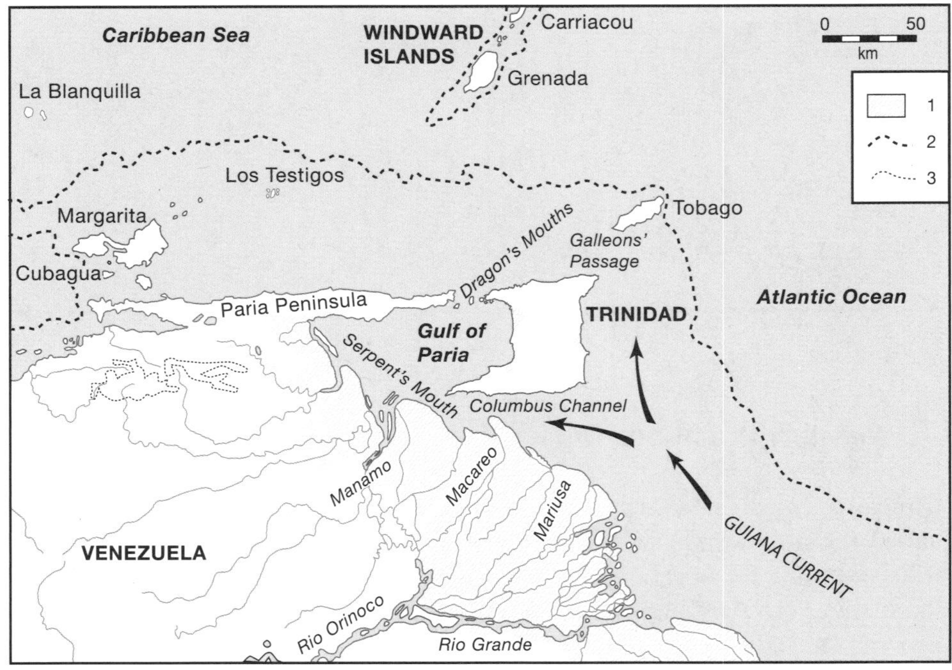

FIGURE 1. The geography of Trinidad. *Legend:* (1) swamps and marshes; (2) edge of continental shelf (200 m below MSL); (3) 1000 m contour.

especially the central part of the island witnesses extensive fluvial flooding because the amount of precipitation exceeds riverine capacities. The island is seldom affected by tropical storms of hurricane force.

The pattern of aquatic circulation around Trinidad is highly affected by the silt-laden South Equatorial Stream, a sea current that moves northwesterly from the South Atlantic along the littoral portion of the Guianas to Trinidad's east and south coasts. In addition, the ecosystem of the waters around the island is determined by the huge amount of muddy water discharged annually by the Orinoco River. Its rainy season outflow even reduces the surface salinity of the Gulf of Paria to such a degree that the condition of the latter approaches that of freshwater, whereas it is near-oceanic in character during the dry portion of the year. Using dugouts, the Warao Indians of the Orinoco Delta easily crossed the Columbus Channel, which separates Trinidad from the mainland (e.g., Wilbert 1993:58–60, 261). Until the 1940s they frequently visited Trinidad for purposes of trade and ritual. It took them twelve to thirteen hours to paddle from the mouth of the Caño Mariusa to the island's south coast. Toward the north, Trinidad is separated from Tobago by the sea channel known as the Galleons' Passage. In the past, Amerindian communication between both islands was facilitated by the circumstance that northeast Trinidad is almost always within view from the southern and central parts of Tobago and vice-versa.

Physiographically, Trinidad consists of five major landform units: three chains of mountains aligned on an east–west axis (the Northern Range, the Central Range, and the Southern Range) with two intervening lowlands (the Northern Basin and the Southern Lowlands) (Figure 2). The Northern Range forms the highest mountain chain of the island, reaching up to 940 m at the Cerro del Aripo. It effectively isolates the northern seashore from the rest of the island. The Northern Range is generally considered the eastern extension of Venezuela's Cordillera de la Costa and consists of strongly upfolded metamorphic rocks, including slatey phyllites, quartzites, and recrystallized limestones dating from Cretaceous through Paleocene times. Steep slopes, caves, springs, and V-shaped valleys characterize its landscape. Relatively fertile soils are to be found in the valley interiors; the mountains are typified by acidic, poorly drained and weathered ultisols. Toward the south, the Northern Range is bounded by the Northern Basin, a depression filled with Pleistocene to Recent sediments, which is drained by the extensive system of Trinidad's largest river, the Caroni. Toward the west, the depression ends in an extensive tidal marsh, covered with mangrove, the Caroni Swamp. The Central Range is a broad highland mass running diagonally across the island, the valleys of which have the best soils of Trinidad. Rounded hill ridges and rugged peaks made up of limestones, siltstones, sandstones, and shales are typical. Toward the west, the Central Range ends in Naparima Hill (179 m), a conspicuously isolated mountain, now forming part of San Fernando. The Southern Lowlands are a structural syncline filled with Miocene and Pliocene sands, silts, clays, and gravels. It is characterized by natural seepages of crude oil, oil-saturated silts and clays ("mud volcanoes"), as well as asphalt. Famous, of course, is Trinidad's Pitch Lake, the most extended asphalt flow of the island (Faber-Morse 2007). The Southern Range, finally, consists of a highly fragmented chain of low hills made up of Miocene sandstones, siltstones, and shales, and typified by variably fertile soils and occurrences of oil sands, mud volcanoes, and asphalt seepages (Boomert 2000:24–30).

Most of Trinidad was formerly covered by tropical rain forest. Unfortunately, human activity has severely affected much of this forest through the monocropping of sugar, cocoa, coffee, and citrus, as well as logging and slash-and-burn cultivation. The mountains of the Northern Range support montane forest, a two-storey type of rain forest with an abundance of ferns, though few lianes and epiphytes. The forest is replaced by elfin woodland on the mountain tops. Central and South Trinidad are typified by evergreen seasonal forest, showing three storeys with many palms in the lower storey. Semi-evergreen and deciduous seasonal forests were formerly quite extensive in the western portion of the island, but most of these formations have disappeared because of the extension of the sugar cane industry. The ecosystems of interior Trinidad also include freshwater swamps and marshes, the latter often showing "islands" of Moriche palms (*Mauritia flexuosa*). Palm marshes and swamp forests are best represented in the Nariva Swamp of the southeast. Herbaceous swamps and natural savannahs are patchily distributed across the island. The savannahs are essentially relics from Pleistocene times.

Trinidad's four coastal stretches are quite different from each other, alternating among sand beaches, rocky shores, and mangroves, as well as other estuarine

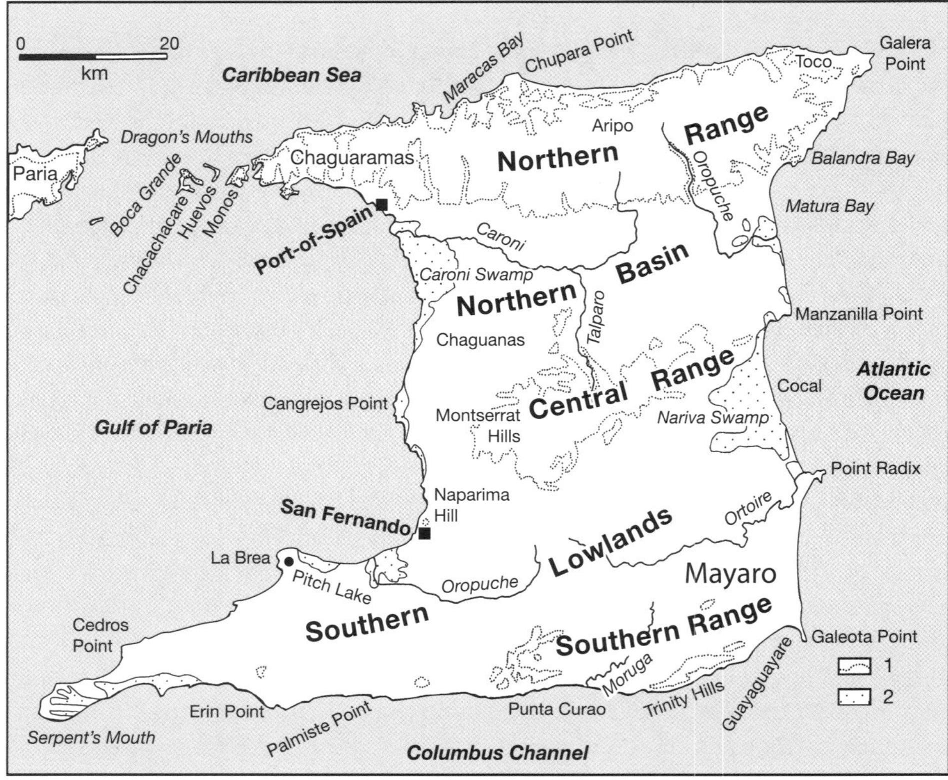

FIGURE 2. The main topographical features of Trinidad. *Legend:* (1) 100 m contour; (2) swamps and marshes.

habitats. Much of the island's littoral grows a characteristically coastal vegetation of dry evergreen forest, also known as littoral woodland, showing stunted, dwarfed, and windblown trees. At present this is the area most suited to coconut cultivation. Coral reefs are encountered only off the northeasternmost part of Trinidad's shore and around the islands in the Dragon's Mouths, the sea channel separating Trinidad from the Paria Peninsula.

Trinidad's natural fauna closely reflects the origin of the island as a piece of South America that separated from the mainland only in Holocene times. Until recently, this habitat, with a highly diverse animal population, closely resembled that of the continent's tropical lowlands. It was much sought after by prehistoric peoples. The upper storeys of the island's rain forest trees are, or were, inhabited by mammals such as anteaters, porcupines, red howler monkeys, capuchin monkeys, tayras, ocelots, and squirrels, while the understoreys were the habitat of opossums, armadillos, and spiny rats, along with various snake species, tree rats, and toads. The forest floors were the domain of red brocket deer, agoutis, pacas, peccaries, ta-pirs, and land tortoises. White-tailed deer, iguanas, and lizards inhabited somewhat more open environments. Typical freshwater swamp dwellers include crab-eating raccoons, otters, capybaras, water rats, anacondas, musk turtles, and caimans. In addition, manatees were formerly common in the rivers of Trinidad's east coast and the Nariva Swamp. The humid edges of swamps form the habitat of numerous land crabs. Freshwater to brackish swamp fish include mullets, snooks, and various species of catfish. Mangroves are typically the spawning grounds of these freshwater and estuarine fishes. Offshore pelagic fish species such as flying fish, tunas, mackerels, requiem sharks, and jacks abound, especially along the east and north coasts. Reef fish such as parrotfishes, grunts, surgeonfishes, snappers, morays, groupers, nurse sharks, and barracudas are common only along the north coast. Numerous edible gastropods are at home in the fresh and brackish water swamp environment, among them tulip mussels, thick lucinas, flat tree oysters and Caribbean oysters. Still highly prized, bivalves such as Donax clams, locally known as *chipchip*, and Trigonal tivelas (*wacoo*) typically inhabit the saturated wash-zone of especially Trinidad's south and east shores. The sand beaches of the northeast are the favored nesting grounds of several sea turtle species, predominantly leatherbacks.

Contact Period Ethnohistory

At the time of the European encounter, Trinidad was a complex multiethnic and multilingual conglomerate of Amerindian groups of possibly differing sociopolitical complexity. In the 1590s and early 1600s at least six indigenous peoples are mentioned as settled on the island—the Chaguana (Warao), Carinepagoto (Carib), Arawak (Lokono), Shebaio, Yaio, and Nepoio—linguistically belonging to three major language families: Cariban, Arawakan, and Waraoan. All of these nations are, or were, found on the South American mainland, notably in the Lower Orinoco River valley, the delta of this river and the coastal zone of the Guianas. Amerindian ethnic patterns show that Trinidad formed a continuation of this portion of the continent, closely reflecting its complicated and segmented structure (Boomert 1984a, 1986, 2000:386, 2004, 2009). By the end of the sixteenth century, the Arawak, Shebaio, and Yaio lived in the coastal area of south and southwest Trinidad, while Nepoio inhabited the central and eastern parts of the island. Trinidad's northwestern portion was occupied by the Carinepagoto, who may have lived in the rest of the Northern Range and along the north coast as well. These groups would have subsisted on horticulture, combined with hunting, fishing, and food collecting. The Chaguana, who in the 1610s inhabited the central coastal area of west Trinidad, may have been an exception, probably being dependent exclusively on fishing and food gathering just as their kinsmen in the Orinoco Delta. Documentary sources dating to the early sixteenth century are less precise than those from the 1590s, referring only to Amerindian groups that can be identified as the Carinepagoto and Arawak. Interestingly, by the mid-sixteenth century the name Arawak was apparently a generic term for many of the Trinidad Amerindians, including the Nepoio, Yaio, and Arawak, who at the time were favorably disposed to the Spanish and traded with the colonists of the pearl islands, Cubagua, and Margarita. It was in this same sense that a century afterward the name Arawak (*Alouagues, Allouäques*) was used by the Island Carib of the Windward Islands.

Clearly, most early-contact Amerindian groups in Trinidad were organized as tribal sociopolitical entities, that is, principally egalitarian societies consisting of at most several thousand people living in a series of semi-independent villages integrated by nonresidential descent groups such as clans, elaborate trade networks, political as well as military alliances, and structures of reciprocal ritual exchange. The individual Amerindian peoples of the island were essentially a series of networks of local groups interconnected by marriages and other exchanges. Most local communities would have been exogamous and apparently consisted of one uxorilocal extended family. By their personal qualities, some headmen were able to dominate in war and exchange, attracting large followings through gift giving. These warchiefs or "great men" had the largest villages because they contracted many marriages and had the right of virilocal residence for themselves and their sons, simultaneously receiving services from their sons-in-law. Having prestige, but not personal power, they had to muster a following by organizing war and trade expeditions and by holding drinking parties and distributing gifts, often exotic goods. Although the position of war chief was not hereditary and theoretically only effective in times of armed conflict, the fact that raiding and trading formed a continual aspect of the social system made his status as a "great man" permanent, at least during his lifetime. Sons of "great men" were also often chosen as their successors, which eventually might have led to the establishment of chiefly lineages. Obviously, long-distance trade and ceremonial exchange next to small-scale warfare formed social institutions that were essential to the functioning of Amerindian societies such as these of early-contact Trinidad.

There are various sixteenth-century assessments of the native population of the island. The first proclaimed full count of the Trinidad Indians, made in 1593, estimated their number at thirty-five thousand, which would be about forty to sixty villages, each counting some five hundred inhabitants (Newson 1976:30). Several documentary sources indicate that the south coast had the highest population density of the island. The description of a dwelling on probably Trinidad's western shore in 1516 suggests that the Indians lived in large, bell-shaped communal houses capable of sheltering up to one hundred people. These houses apparently had double roofs of palm leaves and were closed entirely. A village would have comprised several of these communal houses.

Reacting to the Spanish attempts at colonization of the island in the 1530s, the Carib chief of northwest Trinidad, Baucunar, had his major village, Cumucurapo (today a residential quarter of Port-of-Spain) fortified with defensive walls. Baucunar's "province" is reported to have included at least two villages situated on the Gulf of Paria and others toward the interior. It was ruled by three or four *caciques*, of which Baucunar was the principal one. If so, Baucunar would have been the paramount chief of a "minimal" chiefdom or otherwise a very successful "great man." At any rate, he was able to unite the forces of most of the Trinidad Indians and apparently some groups of the Paria Peninsula against the Spanish. The latter associated themselves with the Trinidad Arawak, who at the time were led by Maluana, a "great man" or chief heading the "province" of Chacomar on the island's south coast (Boomert 1984a, 2000:389–391). During one of the first skirmishes the Spanish killed an Indian "captain," a nephew of Baucunar called "the heir to all of his lands." This is suggestive of matrilineal inheritance.

Complicated patterns of war and exchange characterized the extended interaction sphere encompassing the indigenous societies of Trinidad, Paria, and the Lower Orinoco. Interethnic exchange concentrated on "social valuables," that is, exotic-looking, highly valued and rare manufactures, which changed hands as a means to conclude social and political transactions such as alliances, marriage payments, and homicide compensations. Throughout the Caribbean the major valuables exchanged included *kalikulis*, "Amazon stones," and *quiripá* strings. The *kalikulis* were ear, nose, and breast ornaments in the form of crescents and birds of prey, made from thinly hammered plates of a gold-copper alloy, known as *guanín* by Arawakan speakers. Manufactured in the Colombian Andes, they were transmitted to the east across the Llanos or along the Venezuelan littoral. From the Orinoco the *kalikulis* reached the Island Carib of the West Indies via the Trinidad Arawak, Yaio, and Nepoio, and ultimately arrived among the Taíno in the Greater Antilles. Green animal-shaped pendants, predominantly representing frogs, are known as "Amazon stones" because they were ascribed by the Amerindians to the "women-without-men," a mythical tribe of females that the contact-period European travellers were apt to relate to the Amazons of classical antiquity. These pendants formed the major social valuables of the South American tropical lowlands and were distributed from the Amazon and the coastal part of the Guianas to the Lesser Antilles through the activities of the Kalina and Island Carib. *Quiripá* strings, finally, show a typically western distribution. These were composed of hundreds of small, flat beads of equal size, which were manufactured mainly in the Venezuelan and Colombian Llanos and on the Middle Orinoco. The colonial authorities of Venezuela set up equivalences with Spanish coinage and by the mid-eighteenth century the *quiripá* strings had become the major currency of the region, comparable to the tobacco rolls of the Island Carib in the West Indies.

Background and Methods of Archaeological Research

Until Irving Rouse's first arrival in Trinidad only a handful of archaeological sites on the island had been scientifically examined, all in its southern portion. As early as 1913, Jesse W. Fewkes investigated the Erin settlement site on the southwest coast (Fewkes 1914, 1922:62–78), followed two years later by Theodoor de Booy (1917), who dug at St. Bernard (Mayaro) and Cocal in the southeast. John A. Bullbrook was the first to use modern stratigraphic excavation techniques in Trinidad when he examined the Palo Seco settlement site, close to Erin, in 1919. Unfortunately, his field work at Palo Seco was not published until the 1950s (Bullbrook 1953). Bullbrook continued his research by excavating at Erin from 1934 to 1935 and, under the auspices of the Archaeological Section of the Historical Society of Trinidad and Tobago, in cooperation with Major (then Captain) J. E. L. Carter and Kenneth W. Barr, at this same site from 1941 to 1942 and from 1944 to 1945.

In addition, Bullbrook, Carter, and Barr dug at the Bontour and Quinam settlement sites, also situated in the southern part of the island, in 1942. Following Fewkes, Bullbrook saw Trinidad as an entity unto itself, distinct culturally as well as linguistically from both the South American mainland and the other islands of the Antilles. In fact, he believed that the island was inhabited by only one cultural group throughout most of its prehistoric past (e.g., Bullbrook 1953:68, 90). It was during a short visit to Trinidad in 1941 that Cornelius Osgood of Yale University noted close resemblances between Bullbrook's Erin material and the ceramics he and George D. Howard had just excavated at Los Barrancos on the Lower Orinoco (Osgood 1942; Rouse 1951). Intrigued by these stylistic similarities, Osgood subsequently encouraged Rouse to extend Yale's Caribbean Anthropological Program to Trinidad to further investigate and evaluate these cultural resemblances.

When Rouse arrived in Trinidad, he discussed with Bullbrook his aim at establishing a relative chronological framework for the island, distinguishing the various cultural complexes, or "styles" as he called them, which characterized the island in pre-Columbian times. Bullbrook suggested excavating extensively at only a few settlement and camp sites, including at least one site bearing pottery next to a nonceramic one, in order to get a comprehensive, albeit generalized picture of the Amerindian lifeways in Trinidad during the two major episodes in the island's prehistory. However, Rouse was determined to dig a relatively large number of trenches, distributed over as many sites as possible, to be certain of covering the full range of stylistic variation among the subsequent pottery complexes in Trinidad and to synthesize the latter with the known local sequences of the Orinoco Valley and the West Indies. Also, he was inclined to postpone the investigation of a nonceramic site until after he had established his culture-chronological sequence of Trinidad's pottery complexes.

Cursory examination of the collections available on the island showed Rouse that the remains of three major ceramic complexes were to be found variably at the settlement sites identified by Fewkes, Bullbrook, Carter, and Barr in Trinidad's southwestern portion. Accordingly, Rouse decided to dig at these sites—Palo Seco, Erin, Bontour, and Quinam—afterward using the first three to name the largely subsequent local Trinidadian pottery complexes he was able to distinguish. It was only in the final week of Rouse's first stay in Trinidad that he encountered the Cedros site, which he found to be characteristic of the first ceramic culture of the island (Rouse 1947). When Rouse returned to Trinidad in 1953, accompanied by John M. Goggin, he had as his primary aim to investigate the preceramic era of the island. This was accomplished by excavating at St. John and Ortoire. Goggin's wish to examine Trinidad's late-prehistoric to protohistoric past was fulfilled by working at St. Joseph 2 and Mayo, although at the time it was not understood that the pottery complex typical of the latter site does not represent a continuation of Rouse's Bontour "style."

Rouse's strategy of field research involved excavating one to four trenches into the refuse middens at each site. These trenches were preferably positioned where the deposits seemed to be deepest, enhancing the likelihood of some vertical variation in pottery characteristics. Except in special circumstances, each trench consisted of four to six sections, each forming a 2 by 2 m square. To Rouse this was the smallest size yielding a statistically adequate sample of the available ceramics. Each section was vertically divided into arbitrary 20 cm levels parallel to the surface. A crew of eight men did the digging, working with pick and shovel, as in Rouse's previous Caribbean work. Two men were assigned to a section, one to throw out the soil alongside the trench and the other to search it by hand for artifacts.

Since Rouse's 1946 and 1953 excavations took place during the wet season, showers often interfered with the field work. Fortunately, Bullbrook had advised to bring large tarpaulins so that digging could continue during the rain, but although they kept the excavators dry, the tarpaulins did not prevent water from accumulating in the deeper trenches, and considerable time had to be spent in bailing. In addition, using fine-mesh screens was impossible because of the dampness and stickiness of the soil. The humidity also made it virtually impossible to dry the material found and hampered Rouse's efforts to preserve crushed and broken artifacts with ambroid. Whenever a burial, some structure, or an unusual artifact was encountered, the workmen were stopped and the find dug out with trowel and brush, then charted and photographed to record its exact position. Ordinary specimens were simply bagged by section and level without regard for their situation within these arbitrary units. All man-made finds were preserved, but in some cases sherds that lacked any trace of either shape or decoration were discarded to reduce shipping weight. Similarly, all charcoal and animal bones were bagged, but only a sample of the shell species found in each trench was kept. All sites were mapped and photographed and the vertical cross sections of each trench were recorded. The trenches were dug until sterile soil was reached.

Rouse's 1946 field project in Trinidad resulted in the first relative chronology and prehistoric cultural classification of the island. A preliminary assessment of his excavations at the five southwestern sites led to the formulation of a ceramic sequence consisting of four sequential pottery complexes—Cedros, Palo Seco, Erin, and Bontour (Rouse 1947, fig. 1), of which it proved possible to relate the Cedros and Palo Seco complexes to the Saladoid series (the earliest full horticulturalist culture of the Caribbean). In fact, Cedros was later chosen by Rouse as the type site of the Cedrosan subseries of the Saladoid series because its pottery was representative of the communities belonging to this ceramic tradition in the coastal zone of the mainland and the West Indies as far north as perhaps the eastern tip of Hispaniola (e.g., Rouse 1986, fig. 23). The Erin and Bontour complexes of Trinidad, also identified during the 1946 field season, could be assigned to the Barrancoid and Arauquinoid (Guayabitan) series, respectively, both exclusively represented on the mainland, including the Orinoco Valley and Paria Peninsula. The possibility of ^{14}C dating was not yet available at the time of Rouse's 1946 Trinidad excavation project, but he obtained the first radiocarbon dates for prehistoric Trinidad from charcoal found during his 1953 field work at Ortoire. The animal bone materials, recovered by Rouse (and Goggin) in 1946 and 1953, were studied by Elizabeth S. Wing of the University of Florida at Gainesville (Wing 1962). In fact, this was the first detailed analysis of food remains, apart from shells, encountered at archaeological sites in the Caribbean. It signaled the beginning of a shift in archaeological research from exclusively culture–historical interpretation toward reconstruction of past subsistence patterns and modes of life in the region.

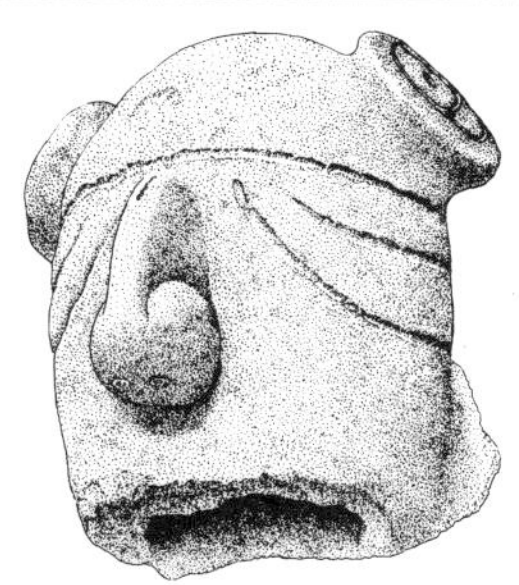

CHAPTER TWO

The Archaeological Sites

The archaeological sites investigated during Yale University's Peabody Museum of Natural History 1946 and 1953 expeditions to Trinidad are discussed here in their approximate chronological order (Figure 3). The system of site classification used follows that of Boomert and Harris (1988), updated by Boomert (2000:495–505). All ^{14}C dates reported are expressed in radiocarbon years Before Present (BP), counted from AD 1950. The quoted errors are one-standard deviation based on the random nature of the radioactive disintegration process. Samples measured before 1962 (the year that AD 1950 was internationally accepted as the BP reference year), are marked with an asterisk. All dates have been calibrated according to the online program CALIB Version 6.0.1 (2010) (Stuiver et al. 2005). Abbreviations used include the following: NAR, Nariva; SGE, St. George; SPA, St. Patrick; VIC, Victoria; YPM ANT, Division of Anthropology, Peabody Museum of Natural History, Yale University, New Haven, Connecticut, USA.

St. John

The Archaic settlement site of St. John (SPA-11), also known as the Oropuche midden, is on top of a 10 m high hillock at the southern edge of the Oropuche Lagoon in Trinidad's Southern Lowlands. It is situated about 875 m south of the Oropuche River, some 3 km inland from Freemans Bay on the Gulf of Paria. A left tributary of the Oropuche River flows at a distance of about 250 m south and west of the site (Figure 4). St. John's Trace, a branch road of St. John's Road, transects the site, formerly exposing a series of shell midden strata. The midden originally measured some 37.5 m in diameter; its former height can be estimated at 1.2 m. At present the site is almost completely destroyed. The midden contained many shells together with cultural materials, rock fragments, flint and chert chips, animal and fish bones, crab claws, and similar remains. It showed a trifold stratification (Figure 5, inset): a bottom zone (Stratum A) mainly composed of Tiger lucinas (*Codakia orbicularis*) and a middle zone (Stratum B) predominantly consisting of nerites (Neritidae), capped by an uppermost layer (Stratum C), including mainly Caribbean oysters (*Crassostrea rhizophorae*) (Harris 1976; Boomert 2000:55–57). The site's natural

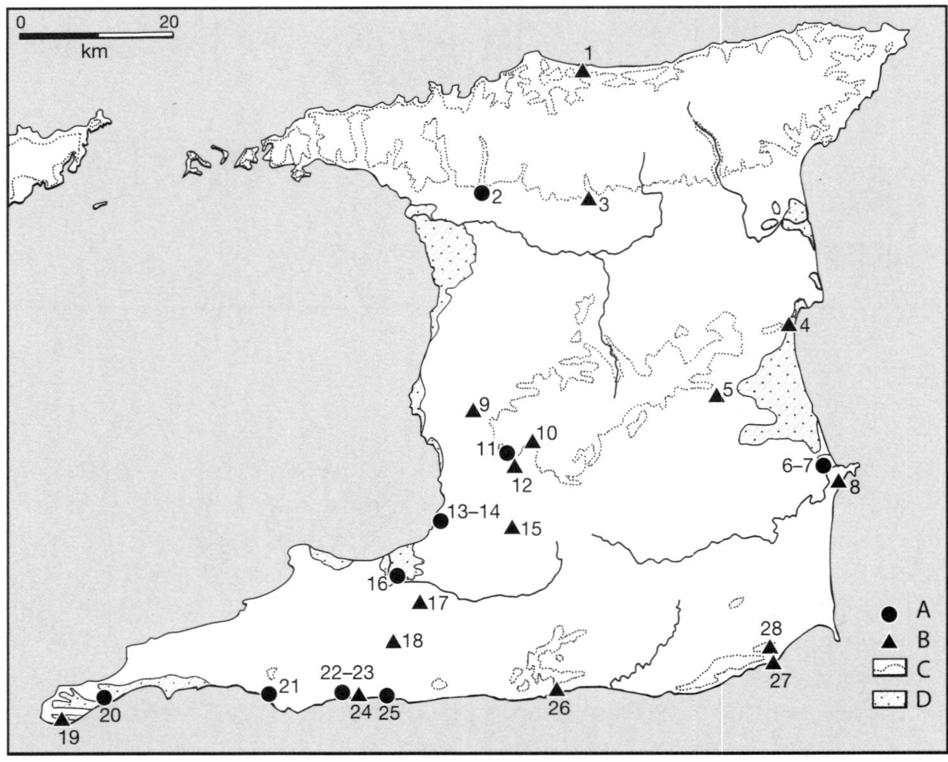

FIGURE 3. The location of selected archaeological sites on Trinidad. *Legend:* A, sites excavated by Rouse in 1946 and 1953. B, other sites mentioned in the text. C, 100 m contour. D, swamps and marshes. *Key to sites:* (1) Blanchisseuse; (2) St. Joseph 2; (3) Arima; (4) Manzanilla 1; (5) Biche; (6) Ortoire; (7) Cocal 1; (8) St. Bernard; (9) Savaneta 2; (10) Atagual; (11) Mayo; (12) Poonah Road; (13) Bontour; (14) San Fernando-Harris Promenade/High Street; (15) Princes Town 1; (16) St. John; (17) Banwari Trace; (18) Siparia-Pastora Street; (19) Icacos; (20) Cedros; (21) Erin; (22) Palo Seco; (23) Palo Seco East; (24) Chagonaray; (25) Quinam; (26) La Lune 1; (27) Guayaguayare; (28) St. Catherine's.

---

vegetation consists of deciduous seasonal forest, but in colonial times this area was used for sugar cane cultivation and formed part of the Belle Vue estate. By the early 1950s it was covered with secondary forest and used as a provision garden; today a house stands on its remaining part. Actively growing mangrove woodland, the Godineau Swamp, is still to be found at less than 1 km from the site. Geologically the site's hillock belongs to the Miocene Springvale Formation. It is characterized by aquentic chromuderts (vertisols) belonging to the Talparo Clay variety, that is, intermediate upland soils with restricted internal drainage. Local precipitation amounts to 150 to 175 cm per year (Figure 6).

The site was discovered by John Bullbrook during a geological survey in 1924. He collected a series of shell samples, which were lost in the 1933 hurricane (Bullbrook 1940:7, 13–18). In 1936 or 1938, oil geologists dug a 5.3 by 0.8 m trench bearing approximately north–south into the portion of the shell midden north

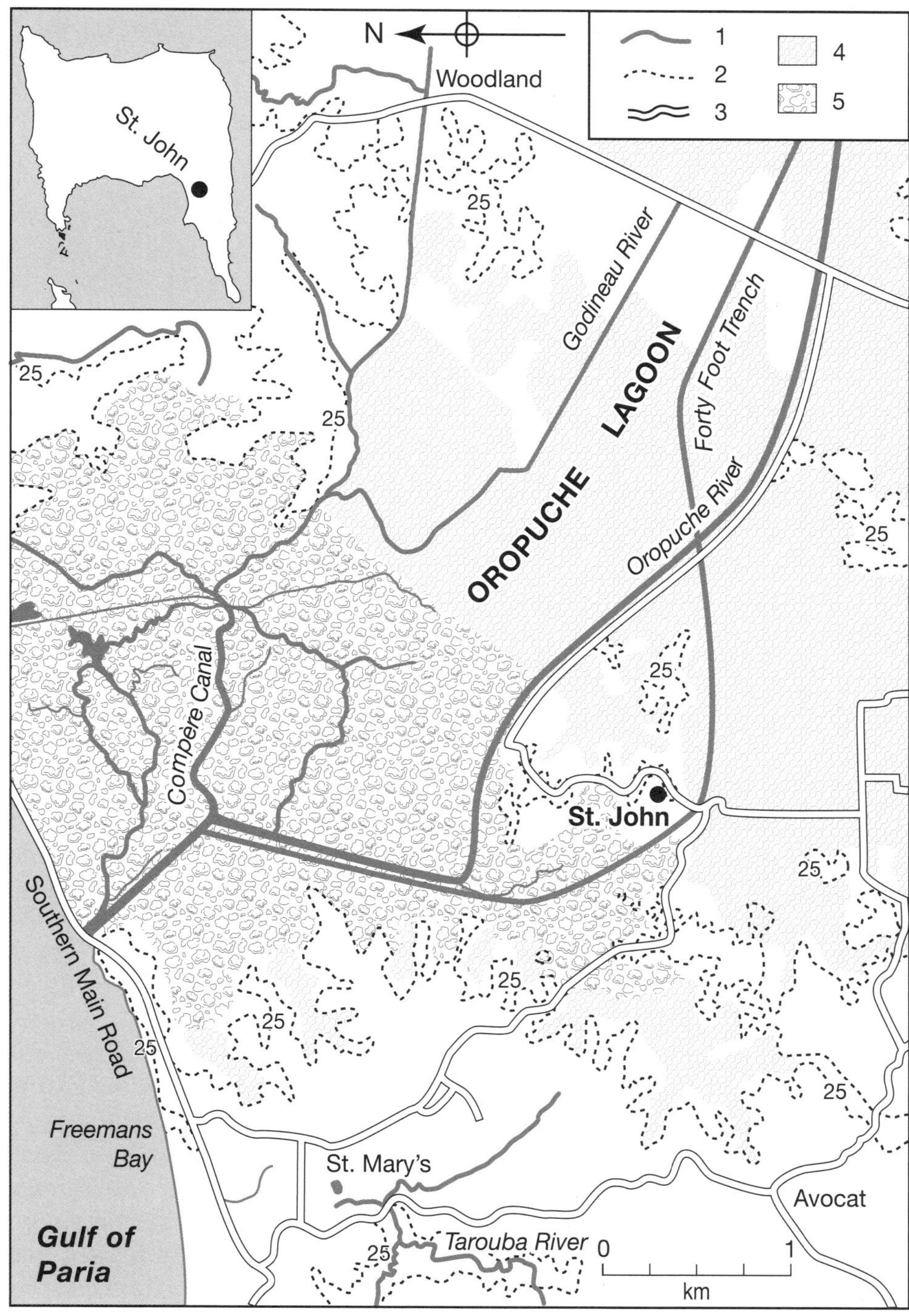

FIGURE 4. The geography of the St. John (SPA-11) archaeological site on the Oropuche Lagoon in southwest Trinidad. *Legend:* (1) rivers and canals; (2) 25- and 100-foot contours; (3) modern roads; (4) swamps and marshes; (5) mangrove woodland.

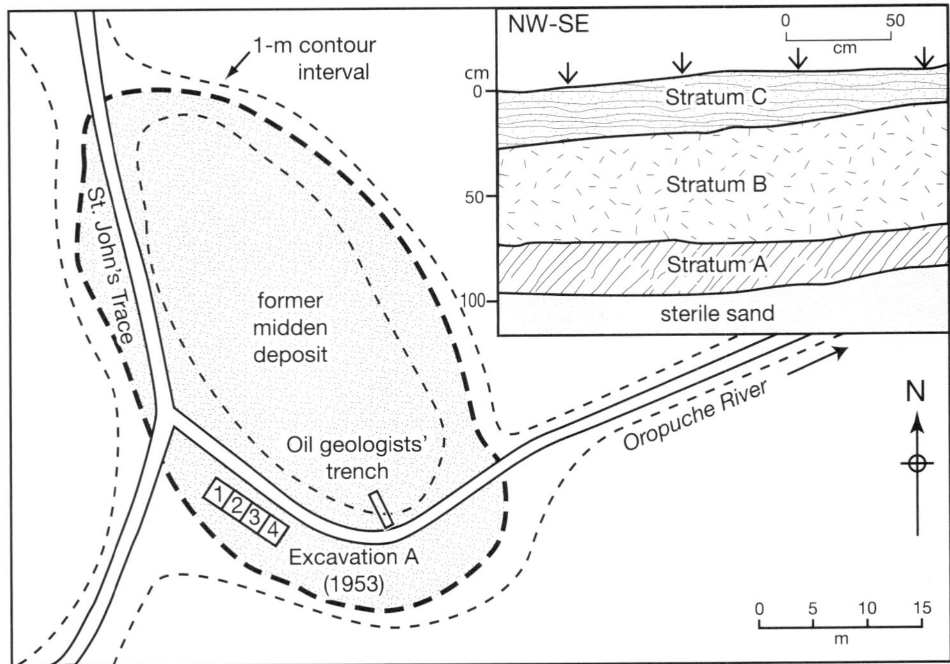

FIGURE 5. The former St. John shell midden site in southwest Trinidad, showing Rouse's 1953 Excavation A and an old oil geologists' trench. *Inset:* Stratigraphic profile of the old trench, recorded by Rouse (for explanation, see text). Adapted from Rouse Papers, St. John field notes.

of St. John's Trace, then a footpath. This trench was still recognizable in the early 1950s (Rouse 1953:101; Bullbrook 1960:10). Because the owner of the site did not allow examination of the part (about 80%) of the midden to the north of the trace, in September 1953 Rouse had to dig south of it, along the southwestern periphery of the midden. His trench (Excavation A), measuring 8 by 2 m, was some 5 m to the southwest of the 1936–1938 excavation, parallel to St. John's Trace. It comprised four 2 by 2 m units (Sections A1 to A4), controlled in 20 cm artificial levels (see Figure 5).

Apart from Archaic cultural materials—notably stone, bone, and shell artifacts of the Banwari Trace complex of the Banwarian subseries, Ortoiroid series (Figures 33 and 34)—Rouse encountered a pottery deposit (105 potsherds) of the Bontour complex, Guayabitan subseries, Arauquinoid series, in the upper layers of the western portion of the midden, mainly in the 0 to 20 cm level, suggesting that St. John had been a camp or bivouac site in late prehistoric times. Rouse also found historic period materials, such as English ceramics, bricks, glass, and iron nails dating from the early nineteenth century (Figures 82 and 84) in this same part of the site and similarly predominantly in the midden's upper excavation level in apparent mechanical admixture with the Bontour earthenware (Rouse Papers, St. John field notes). This stratigraphic situation has given rise to the erroneous conclusion that the Bontour complex could perhaps be dated as late as the nineteenth century (Rouse 1953:96, 101, 108, 110; Wing 1962:33–35). Sterile, sandy soil was

Figure 6. View of the Oropuche Swamp in 2004, directly west of the former St. John archaeological site in southwest Trinidad.

reached at a depth of 70 cm below the present surface in Rouse's Section A1 and at 60 cm depth in Sections A2 to A4. Shell Stratum C, clearly distinguishable in the 1936–1938 trench, was not encountered in Rouse's excavation (Harris 1971:7, 1976, fig. 6). In fact, the two upper levels of his trench contained much charcoal and ash. All the finds excavated by Rouse at St. John were deposited at the Yale Peabody Museum of Natural History. Rouse's zoological remains were analyzed by Wing (1962:33–38, 1977; Wing and Reitz 1982; see also Newson 1976:48, 51–54; Boomert 2000:62–64).

After Rouse's investigations, the site's stratification was inspected by José M. Cruxent (Caracas) and Peter O'Brien Harris in September 1969 (TTHS Annual Report 1969). The latter revisited the site in November–December 1970 and found that by far most of the remaining part of the midden had been destroyed in that year, by being removed to provide garden topsoil for private lawns in San Fernando and Valsayn. The following year members of the Trinidad and Tobago Historical Society (South Section), led by Harris, paid a series of visits to the St. John site, rescuing artifacts from the surface. In July and October–November 1972, Harris dug a 7 by 1 m radial trench (Excavation B), consisting of seven excavation units (Sections B1 to B7), from the estimated center of the midden to its perimeter (Harris, St. John field notes). He encountered a variety of stone, bone, and shell artifacts dating from Archaic times, as well as a hearth feature consisting of a sand bed with a clay center (Harris 1971:33, 1973, 1974, 1976; TTHS Annual Report 1971; TTHS

Newsletter 1971(1, 9), 1972(4), 1973(1)). Harris revisited St. John in January 1975, July 1977, March 1978, and February–May 1979, each time collecting surface materials (TTHS Annual Report 1977–1978; TTHS Newsletter 1978(3, 4), 1979(2, 5)). All finds encountered by Harris are in the collections of the University of the West Indies in St. Augustine, Trinidad, and the Pointe-à-Pierre Wildfowl Trust, Petrotrin, Ltd., in Pointe-à-Pierre, Trinidad.

Subsequently, the St. John site was visited by Archibald S. Chauharjasingh and Nicholas J. Saunders in October 1992 and again in May 1997 (Saunders and Chauharjasingh 2003:40–42). They collected some cultural and archaeozoological materials from the surface of the site, which are now kept at the University of the West Indies. Chauharjasingh donated a sample (S.JO/01) of marine shells, including a West Indian Crown conch (*Melongena melongena*) found by Harris at an unspecified depth in Excavation B in November–December 1972, to François Rodriguez-Loubet (Martinique). The latter obtained a radiocarbon date (ARC-1153) for this sample of 6866±48 BP, or 5475–5378 cal BC (Boomert 2000:57, 516), suggesting that it was associated with materials that can be assigned to the Early Banwari Trace complex, also known from the Banwari Trace site (SPA-28), some 5 km farther inland on the southern edge of the Oropuche Lagoon. Finally, the site's present conditions were inspected by Boomert and Harris in August 2004.

Ortoire

The Archaic settlement site of Ortoire (NAR-4) is situated on flat land under coconut cultivation to the east of a mangrove swamp at the southwestern edge of the Nariva Swamp. The site is about 100 m west of the Manzanilla Mayaro Road, some 175 m inland from Cocos Bay, and about 1 km from the mouth of the Ortoire River. A small freshwater stream empties into the sea directly north of the site, which formerly consisted of a single shell midden that stood out conspicuously in the landscape (Figure 7). Originally it measured 23 by 14 m, with a maximum height of 1.25 m (Rouse Papers, Ortoire field notes). In the early 1950s only the southeastern edge of the midden had been damaged (by the removal of shells for road gravelling). In the 1980s its entire center and large parts of its northern and eastern flanks were also removed for this purpose and only those areas that supported coconut trees were relatively undisturbed (TTHS Newsletter 1972(4); Boomert 1984b:57–58, 1987:38–39). The midden is now completely destroyed. It consisted of mainly marine shells mixed with dark brown sandy soil. Ortoire is the type site of the Ortoire complex of the Ortoiran subseries, Ortoiroid series, and has yielded stone and bone artifacts along with stone fragments and flint chips of this complex (Figures 35 and 36), as well as animal and fish bones, crab claws, and charred nuts. The Ortoire site is about 575 m south–southeast of another Archaic shell midden deposit, Cocal 1 (NAR-3), which was excavated by Theodoor de Booy as early as 1915 (de Booy 1917). This site, consisting predominantly of Trigonal tivelas (*Tivela*

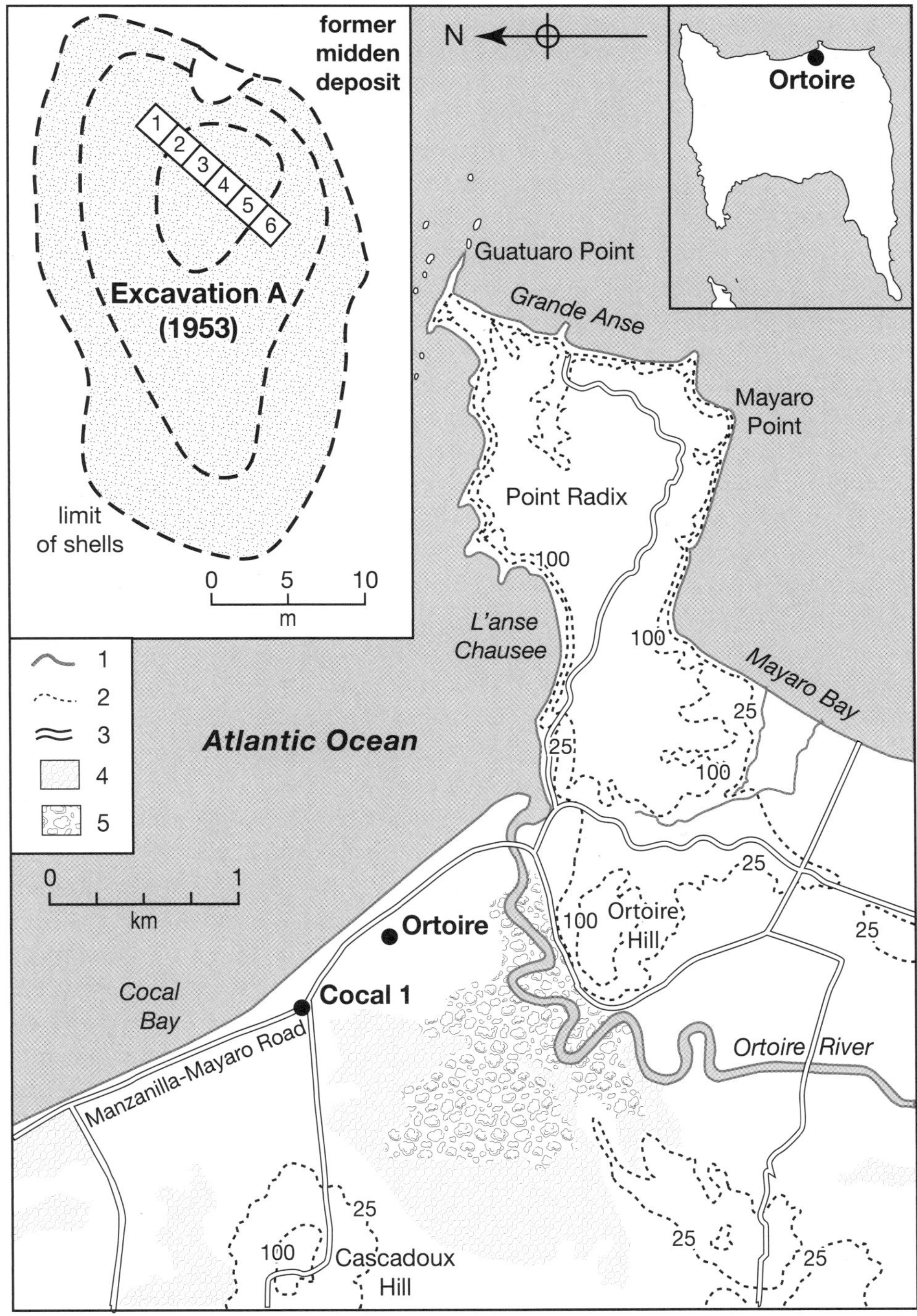

FIGURE 7. The geography of the Ortoire (NAR-4) and Cocal 1 (NAR-3) archaeological sites in southeast Trinidad. *Legend:* (1) rivers and canals; (2) 25- and 100-foot contours; (3) modern roads; (4) swamps and marshes; (5) mangrove woodland. *Inset:* Location of Rouse's 1953 Excavation A at the former Ortoire shell midden site. Adapted from Rouse Papers, Ortoire field notes.

mactroides), locally known as *wacoo*, is considered to belong to the Ortoire complex as well (Boomert 1984b:52–54, 2000:84–85). Both sites are on the north–south sandbar (chenier) separating the Nariva Swamp from the Atlantic Ocean. This sandbar originally supported semi-evergreen seasonal forest. Pedologically, this area belongs to the typic quartzipsamments (entisols) of the Cocal Sands variety. Local precipitation is 175 to 200 cm per year.

The Ortoire site was pointed out to Rouse and the Goggins by a local hunter when they were visiting the Cocal 1 midden in August 1953. Subsequently, Rouse dug a southwest–northeast trench (Excavation A) across the highest point of the site, in the southeastern portion of the midden. This trench, which measured 12 by 2 m, was divided into six 2 by 2 m excavation units (Sections A1 to A6), controlled in 20 cm artificial levels (Figure 7, inset). Sterile yellow sand was reached after digging six to seven levels, at a depth of 110 to 135 cm below the modern surface of the midden (Rouse 1953:96, 106). Apart from shells and other food remains next to cultural remains (Rouse 1960:10–11; Boomert 2000:86), Rouse encountered several patches of clay associated with ash, fire-cracked stones, burned shells, and substantial amounts of charcoal, most likely remains of hearths. One of these clay lenses had a diameter of 55 cm and a thickness of 25 cm (Rouse Papers, Ortoire field notes). The Ortoire midden appeared to be composed of basically three strata: a lowermost zone (Stratum I) between 90–120 and 135 cm below the present surface, composed of predominantly Donax clams (*Donax* spp.), locally known as *chipchip*, and a few Trigonal tivelas, capped by a central zone (Stratum II) between a depth of 30 and 90–120 cm (thus making up most of the midden) characterized by mainly Trigonal tivelas, and an uppermost layer (Stratum III) to a depth of 30 cm below the present surface yielding broken shells of predominantly this same species (Harris 1976, fig. 7). The stratification of the remaining parts of the site was studied by Harris in June 1972 and May 1974 (TTHS Newsletter 1972(4); Harris, Ortoire field notes, 1976), followed by Harris and Boomert in May 1984. The site was destroyed about 1989 when the estate owner had the coconut trees of the area cleared to prepare beds for growing watermelon and cassava. The terrain was ploughed and the remaining parts of the midden leveled. Secondarily worked shell debris can still be seen on the surface at the site (Peter O'Brien Harris, pers. comm. to Boomert 2005). A sample of charcoal retrieved from "all sections" (Sample O-A-5) of Rouse's Excavation A, collected from 80 to 100 cm below the present surface (from the transition between Strata I and II) yielded a radiocarbon date (Y-260-2) of 2760±130 BP*, pointing to a corrected date of 1112–798 cal BC. Two combined charcoal samples (Samples O-A-6/7) from "the bottom of" levels 6 and 7 (100 to 140 cm depth) of "all sections" (Stratum I) yielded a date (Y-260-1) of 2750±130 BP*, pointing to a corrected date of 1111–794 cal BC. A duplicate measurement that produced a date of 2420±140 BP* is considered to be less reliable (Boomert 1984b:58–59, 2000:516). All Rouse's finds are in the Yale Peabody Musuem, and those collected by Harris and Boomert are at the University of the West Indies.

Cedros

The Ceramic settlement site of Cedros (SPA-1) is the type site of the Cedros complex of the Cedrosan subseries, Saladoid series. It is in a former coconut plantation at the foot of the southeastern slope of Green Hill, formerly known as Coffee Hill, an eminence (54.3 m) some 500 m inland from the shore of the Columbus Channel on the Cedros Peninsula of southwest Trinidad (Figure 8). The site, a series of shell midden deposits about 200 m inland from the sea, occupies a low rise forming part of a gently sloping ridge extending diagonally from Green Hill to the southeast. A footpath, which once connected this hill directly with the Columbus Channel, passed through the eastern part of the site. The Cedros site, which measures at least 0.85 ha, occupies a coconut grove belonging to the former Beaulieu estate on both sides of St. Quintin Road East, between approximately 235 and 415 m west of its junction with the Beaulieu Road (Figure 9). The latter road forms a left branch of the St. Marie Road, which crosses the Cedros Peninsula from Bonasse to Galfa Point.

The Beaulieu estate was acquired by the government of Trinidad and Tobago in the 1970s. Subsequently, it was divided into lots that were given out for use as agricultural plots to the inhabitants of Bonasse. The Cedros site occupies several of these lots, some of which are currently uncultivated. A small, unnamed (slightly brackish) stream flows along the base of the hill in the northeastern portion of the site (Rouse 1953:96, Papers, Cedros field notes; Duymelinck and Bright 2005; see also Wing 1962:38; Olsen 1969a, 1973, 1974:245, 255–256, map 13; Boomert 2000:129; Chauharjasingh, pers. comm. to Boomert 2005). This area was once covered by evergreen seasonal forest. An extended freshwater swamp and mangrove woodland, the Los Blanquizales Lagoon, is less than 2 km to the east of the site, just across the Beaulieu Road, while various mud volcano deposits are situated directly northwest and east. Landslides due to the motion of mudflows are rampant in the area, which pedologically is dominated by vertic ustropepts (inceptisols) of the Green Hill variety that developed from the calcareous volcanic mudflow deposits. Local rainfall amounts to 125 to 150 cm per year.

The Cedros site was discovered by oil geologists as early as 1923. The following year H. G. Kugler dug a geological test pit at the site and encountered the shell midden deposit. Subsequently, local people and a priest from St. Lucia, Father Guilbert, visited the site, collecting surface materials, while Bullbrook inspected it in 1945. Some potsherds and chert chips, kept in the National Museum of Trinidad and Tobago in Port-of-Spain, may derive from Kugler's activities at Cedros. In August 1946, Rouse first excavated a 1 m diameter test pit at the Cedros site, encountering a shell midden deposit of predominantly Donax clams (*Donax* spp.), reaching sterile, yellowish soil at a depth of 80 to 120 cm. Rouse later dug a northwest–southeast 12 by 2 m trench (Figure 10, Excavation A) between two rows of coconut palms just east of the center of the midden and extending from St. Quintin Road East toward the sea. This trench, which consisted of six 2 by 2 m units (Sec-

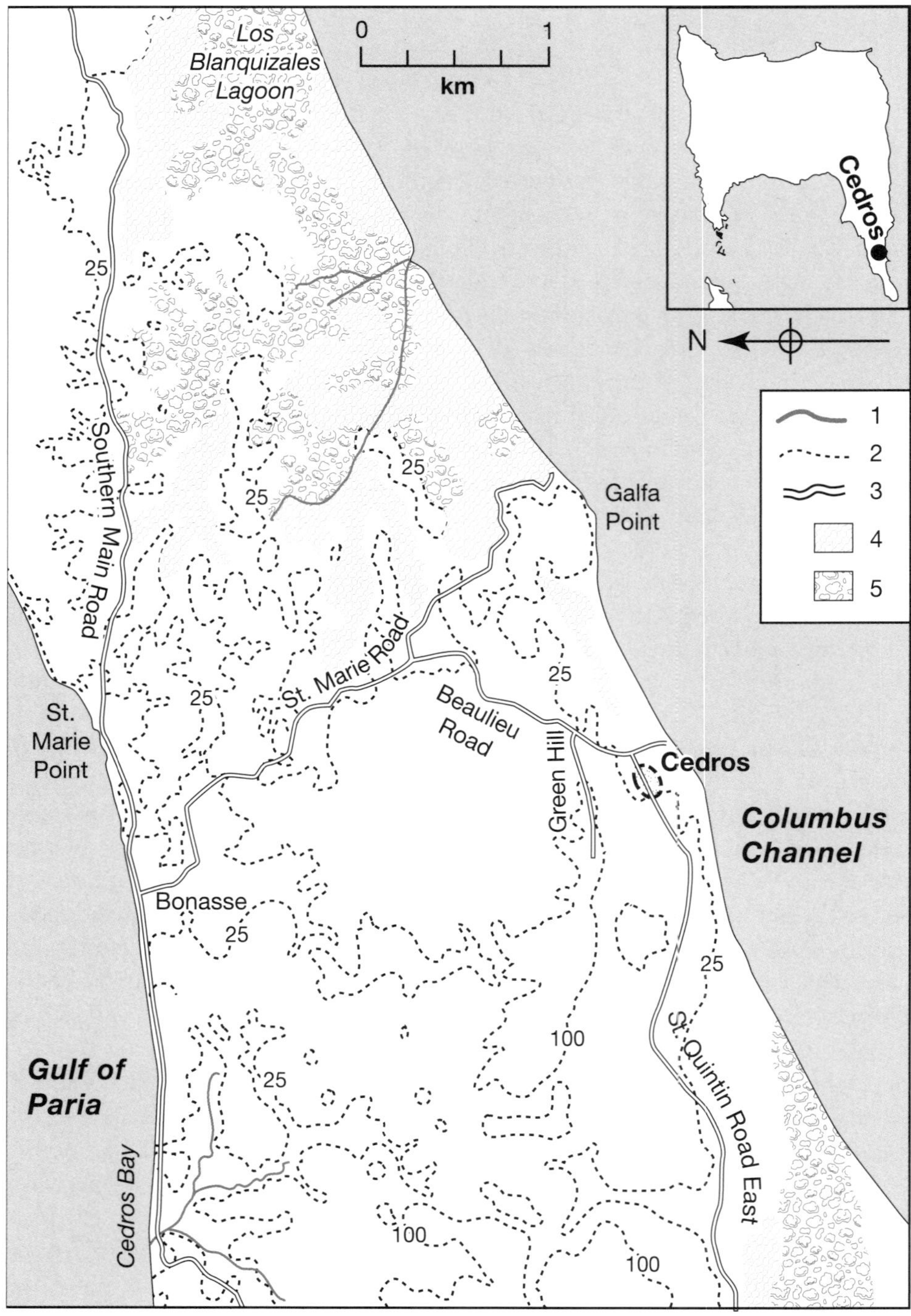

FIGURE 8. The geography of the Cedros (SPA-1) archaeological site in southwest Trinidad. *Legend:* (1) rivers; (2) 25- and 100-foot contours; (3) modern roads; (4) swamps and marshes; (5) mangrove woodland.

FIGURE 9. View of the Cedros archaeological site in southwest Trinidad in 2005. Photograph courtesy of Alistair J. Bright.

tions A1 to A6), controlled in 20 cm artificial levels, yielded refuse until a depth of 60 to 70 cm below the surface. The deposit was of dark brown sandy loam associated with shells, pottery and stone, bone and pottery artifacts, along with animal bones, stone fragments, and charcoal, resting on sterile yellow clay. A piece of a burned human long bone was recovered from the 60 to 80 cm level of Section A6. The pottery can be assigned to the Cedros complex (Figures 42–45); the top levels of the site (to a depth of 40 cm) yielded some material that may be transitional to the Palo Seco complex of the Cedrosan subseries, Saladoid series (Rouse 1947, 1953:102–105, Papers, Cedros field notes; also Olsen 1973, 1974:238–239; Kyberg 1976; Boomert 2000:131–145; Faber-Morse 2007). All Rouse's finds are in the Yale Peabody Museum (YPM ANT 166200–168782).

Elizabeth S. Wing analyzed the animal bone materials encountered by Rouse during his work at Cedros. In addition, she studied a small zoological collection made at the site by Kugler and some surface finds she discovered herself during a visit to the site in August 1959. The latter are curated at the Florida State Museum at the University of Florida, Gainesville (Wing 1962:4, 38–39, 46, 49–51, 1977; see also Newson 1976:48, 52–53; Boomert 2000:340). Rouse, Fred Olsen, and José M. Cruxent visited the site in September 1969 to obtain samples for radiocarbon dating. Assisted by Harris, John A. Correia, and R. Douglas Archibald, they excavated some "exploratory test pits" and a 2 by 2 m square (Pit 1), controlled in 25 cm arti-

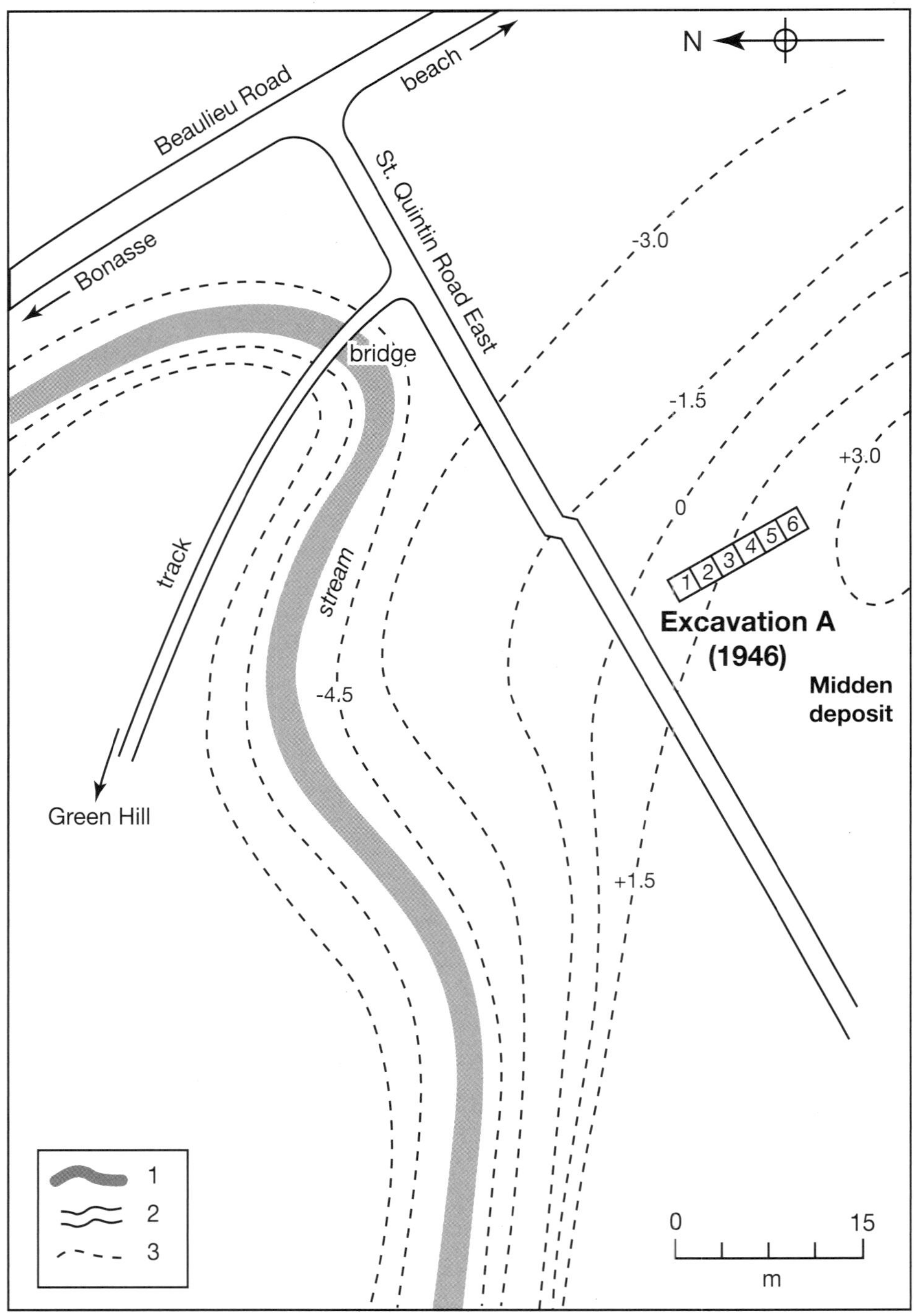

FIGURE 10. The Cedros shell midden site in southwest Trinidad, showing Rouse's 1946 Excavation A. *Legend:* (1) stream; (2) modern roads and tracks; (3) 1.5 m contours. Adapted from Rouse Papers, Cedros field notes.

ficial levels, some 12 m south of St. Quintin Road East and 256 m from its junction with the Beaulieu Road, "only a few feet away" from Rouse's 1946 trench. They encountered shells, bones, Cedros complex ceramics, and unworked rock fragments, including a piece of amethyst, at a depth of 70 cm below the modern surface, resting on sterile yellowish soil.

The combined charcoal bits from the 25 to 50 cm level (the "intermediate layer") of this pit yielded a radiocarbon date (IVIC-642) of 2140±70 BP, or 352–356 cal BC, while that from the 50 to 75 cm level (the "lowest layer") returned a date (IVIC-643) of 1850±80 BP, i.e., cal AD 69–251 (Boomert 2000:518). The charcoal "was in granules about the size of a grain of wheat or smaller" (Olsen 1969a, 1973, 1974:247). Although these dates are not in agreement with the stratigraphic order in which the samples were collected, the oldest one is quite acceptable for the Cedros complex. Unfortunately, the archaeological specimens collected from Pit 1 were lost during shipment to Venezuela (TTHS Annual Report 1969; Archibald 1971:20; Olsen 1973, 1974:245–247; Boomert 2000:10). The Cedros site was visited in March 1993 and April 1995 by Saunders and Chauharjasingh, the second time accompanied by Alex de Verteuil (Saunders and Chauharjasingh 2003:33; Chauharjasingh, pers. comm. to Boomert 2005).

The site's present condition was inspected by Boomert and Harris in August 2004. In July and August 2005 a four-week excavation project by Boomert and Alistair J. Bright was undertaken at the Cedros site as a joint project of Leiden University, The Netherlands, the University of the West Indies, and the Trinidad and Tobago Archaeological Volunteers' Group led by Harris (Duymelinck and Bright 2005). The site was surveyed taking the junction of the Beaulieu Road and St. Quintin Road East as the base point. The extension of the various shell midden deposits was established by investigating a series of ninety-one auger holes, distributed according to a measuring grid system laid out over the site. As orientation for the grid's baseline, St. Quintin Road East was chosen. The auger rows were spaced 20 to 25 m apart, stretching perpendicularly from the baseline through the bush toward the sea in one direction and toward Green Hill in the opposite direction. The auger points were spaced 10 m apart whenever possible. The auger data suggest that the site consists of a continuous band of midden deposit stretching some 110 m and an isolated midden in the extreme east, running the length of St. Quintin Road East. There is a suggestion of a horseshoe-shaped midden configuration, with either end curving south toward the sea, but this may be an artifact of sampling as only part of the terrain has been augered thus far and several auger points set out by hand in the "open" southern part of the horseshoe also revealed the presence of shell deposits (Bright 2006).

In addition, two adjoining 1 by 1 m test pits (B1 and B2) excavated in a portion of the site, 25 m south of St. Quintin Road East, showed a dense concentration of shells on the surface. Both pits were controlled in 5 cm artificial levels kept parallel to the surface. The excavated soil was wet-screened in the Columbus Channel, using fine-mesh (2 mm) sieves. Only test pit B2 was dug through the entire shell

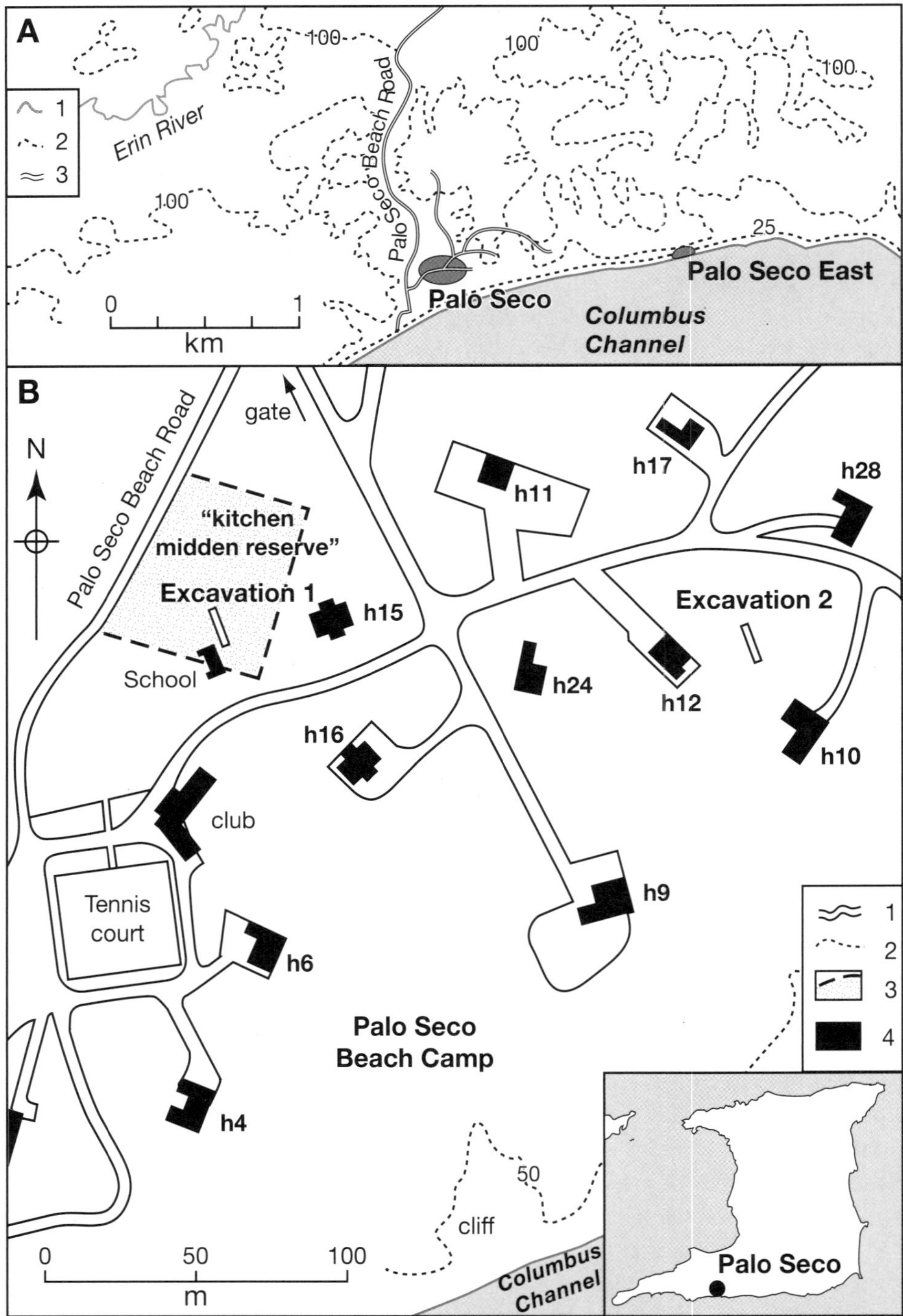

FIGURE 11. Palo Seco sites. **A,** The geography of the Palo Seco (SPA-30) and Palo Seco East (SPA-8) archaeological sites in southwest Trinidad. *Legend:* (1) rivers; (2) 25- and 100-foot contours; (3) modern roads. **B,** The Palo Seco shell midden site, showing Rouse's 1946 Excavations 1 and 2. *Legend:* (1) modern roads and tracks; (2) 50-foot contour; (3) boundary of Palo Seco kitchen midden reserve; (4) buildings (houses are numbered). Partially adapted from Rouse Papers, Palo Seco field notes.

midden deposit. It yielded loose shells mixed with dark brown sandy clay, at a depth of 20 to 30 cm below the present surface turning to a dense accumulation of shells, predominantly Donax clams, which rested on yellowish sterile sand at a depth of 80 to 95 cm below the modern surface. This shell deposit was found to contain pottery dating to the transitional period between the Cedros and Palo Seco complexes, stone artifacts, charcoal, charred nuts, and animal and fish remains, along with stone fragments and shell debris.

All finds retrieved during the 2005 Cedros excavations are kept at the depot of the Trinidad and Tobago Archaeological Volunteers' Group; analysis of the materials is in progress. Samples of charcoal collected from levels 15 to 17 (70 to 85 cm depth), levels 13 and 14 (60 to 70 cm depth), and levels 11 and 12 (50 to 60 cm depth) of test pit B2 yielded radiocarbon dates of 1780±30 BP (GrN-30,418), or cal AD 181–324, 1730±30 BP (GrN-30,419), or cal AD 255–375, and 1570±30 BP (GrN-30,420), or cal AD 434–537, respectively.

Palo Seco

The multicomponent Ceramic settlement site of Palo Seco (SPA-30) is the type site of the Palo Seco complex of the Cedrosan subseries, Saladoid series. It occupies about half of the Palo Seco Beach Camp, owned by Petrotrin, Ltd., and is situated at the end of the Palo Seco Beach Road, some 3.5 km south of the village of Palo Seco on the San Fernando Siparia Erin Road, southwest of Los Bajos. The site is some 200 to 250 m inland from the shore of the Columbus Channel (Figure 11A). It consists of a series of distinct, more or less oval shell midden deposits occupying the top and flanks of a gently sloping, about 25 m high ridge that was originally covered with semi-evergreen seasonal forest. This ridge rises from the cliff edge to a sharp hill of greater height beyond the refuse deposits. Midden 1 forms the main concentration, showing a northeast–southwest axis and measuring some 55 by 30 m, although shell refuse has been encountered here covering an area of in all about 150 by 100 m (Bullbrook 1953:17–23, fig. 2). It more or less fills a shallow hollow in the original hillside, therefore no heaping is seen on the surface. This midden, which faces away from the sea, occupies most of the local schoolyard, just north of the Palo Seco Beach Club (Figure 12). A lesser deposit, Midden 2, is about 180 m to the east, between some bungalows. It primarily occupies the gentle southeastern flank of the hill, facing the sea (Figure 13). Various other, reportedly thinner refuse deposits, facing the sea, have been identified between the two major deposits, for example, southwest of the club house and the adjoining tennis court. The source of potable water nearest to the site is a spring about 750 m east of the site. A perennial stream, the Erin River, flows some 1.5 km to the northwest (Bullbrook 1940:13, 1953:8, 15, 17, 21, 57, 1960:17; Boomert 2000:129–130; Rouse Papers, Palo Seco field notes). Geologically, the site is characterized by the Miocene clays of the Upper Cruse Formation. Soils are dominated by entic pelluderts (vertisols) of the Cromarty Clay variety. Precipitation amounts to 125 to 150 cm per year.

FIGURE 12. View of Palo Seco archaeological site in southwest Trinidad, Midden 1, in 2004.

The Palo Seco site was discovered by E. Cunningham Craig, the then government geologist, in 1906. However, Craig interpreted the shell deposit as a natural phenomenon and it was not until February 1917 that it was recognized by T. Hitchon as a prehistoric Amerindian settlement site. At the time, the site was being dug for shells used for road gravelling by the colony's Public Works Department. Hitchon warned Bullbrook, who succeeded in convincing the then governor of Trinidad, Sir John Chancellor, to have this stopped. Subsequently, Chancellor asked Bullbrook to excavate the site, at government expense, and had part of it demarcated as a Crown Reserve, which it still is. Delayed by the circumstances of World War I, Bullbrook could not undertake the investigation until 1919, when the excavations from April to September took place under the auspices of the British Museum, London (Bullbrook 1953:8, 15).

Bullbrook dug two crossing trenches in Midden 1, Trenches A and B–C, measuring 53 by 1.5 m and 24 by 1.5 m, respectively, and encountered shell refuse accompanied by archaeological remains to a depth of at most 150 cm below the present surface. According to Bullbrook (1940:10–13, 1953:21–29, 26, 92, figs. 2 and 3, 1960:31, 1961b:2–3), the vertical cross section of Trench A yielded a stratification of two layers, characterized by shell refuse deposits of quite different contents and showing "complete unconformity." Below a top layer of dark brown humus he encountered an "upper stratum" of dark brown loam yielding predominantly Donax

FIGURE 13. View of Palo Seco archaeological site in southwest Trinidad, Midden 2, in 2004.

clams (*Donax* spp.), which surmounted a "lower stratum" consisting of medium brown sandy clay with mainly Trigonal tivelas (*Tivela mactroides*) accompanied by smaller quantities of Donax clams and nerites (Neritinae). The lower stratum, which showed various clay lenticles interrupting the shell deposit, rested on sterile, yellowish clay. While Bullbrook's lower stratum yielded pottery, stone, bone, and shell artifacts of exclusively the Palo Seco complex, the upper stratum contained Palo Seco materials next to ceramics of the Erin complex, Barrancoid series, and a few specimens of the Bontour complex, Guayabitan subseries, Arauquinoid series. Bullbrook estimated that a considerable time interval existed between the two subsequent periods of refuse deposition (Bullbrook 1940:10–13, 1953:23–26, 68, 73, 1960:31; Rouse 1947, fig. 1, 1953:34, 82–83, 90; Boomert 2000:357).

In all, eleven human burials were recovered by Bullbrook during his Palo Seco excavations. All skeletons except one were interred in strongly flexed positions, but the orientation differed (Bullbrook 1927, 1940, 1949, 1953:17, 21, 57, 1960:17). Only two burials, both found in the lower stratum, yielded mortuary gifts. One of these, a flexed primary interment of an individual adult, probably male, yielded a necked jar, a pottery disc, and a flanged bowl. Another, a tall male skeleton, lying extended on his back with the head toward the north, was provided with a comparably flanged bowl and a stone axe head placed between the legs. A third burial, unaccompanied by grave offerings, was similarly encountered in the

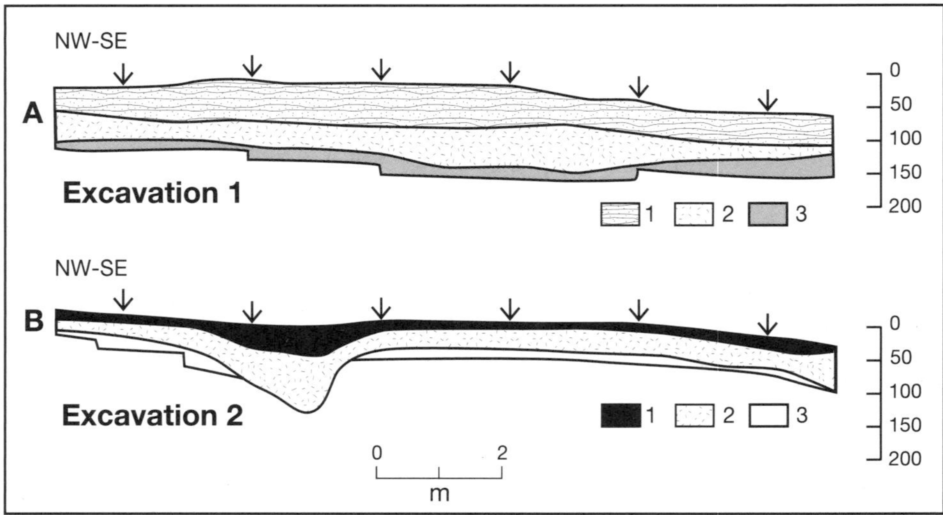

FIGURE 14. Stratigraphic profiles of Rouse's Palo Seco Excavations 1 and 2 (for explanation, see text). **A,** Excavation 1. *Legend:* (1) dark brown loam, many shells; (2) medium brown clay, moderate shells; (3) light brown clay, sterile. **B,** Excavation 2. *Legend:* (1) black humus, few shells; (2) medium brown clay, moderate shells; (3) yellow clay, sterile. Depth is in centimeters. Adapted from Rouse Papers, Palo Seco field notes.

lower stratum. Four inhumations were found associated with the upper stratum. Finally, Bullbrook recovered three additional burials from a disturbed section of the site, probably belonging to the upper stratum, and another away from the midden "in solid clay" (Bullbrook 1940:40, 1953:21, 39–40, 57, 1960:17, 49, 71; Boomert 2000:158, 399–400). In addition, hearths (more or less oval, pink areas containing ash, charcoal, and burned shells) were found, especially in the upper stratum. Charcoal seemed to be abundant in the bottom layer of the deposit (Bullbrook 1940:13, 1953:26, 28–29, 67, 1961b:2), possibly a result of the forest clearance operations before the construction of the settlement. Apart from pottery remains, Trenches A and B–C yielded various artifacts of bone, stone, and pottery, as well as unworked pieces of stone and shell next to zoological remains (Bullbrook 1953:339–344). All Bullbrook's 1919 finds are in the British Museum, London (see Boomert 2000:9). Some pottery remains from the Palo Seco site are curated in the National Museum of Trinidad and Tobago. Most likely these were collected at the site by C. C. Winkler about 1940.

In July 1946, Rouse dug a northwest–southeast trench (Excavation 1) in Midden 1, at a distance of about 2 m stretching "alongside" Bullbrook's Trench A, some 6 m north of the camp primary school. This trench, measuring 12 by 2 m, consisted of six 2 by 2 m units (Sections A1 to A6), controlled in 20 cm artificial levels (Figure 11B). Section A1 was closest to the sea; the other units extended uphill in the direction of Bullbrook's Trench B–C. The latter's bifold stratification was distinct only in Sections A3 to A6. These units yielded dark brown loam or humus and shell refuse of Bullbrook's upper stratum until a depth of 40 to 60 cm below the modern surface, superimposed on a layer of medium brown sandy loam and shells of Bullbrook's lower stratum, turning into light brown to yellowish sterile clay at a depth of 120 to 140 cm (Figure 14A). Charcoal was found throughout (Rouse Papers, Palo Seco field notes). A second 12 by 2 m northwest–southeast trench (Excavation 2) was dug by Rouse in Midden 2. This trench similarly consisted of six 2 by 2 m units (Sections G1 to G6), controlled in 20 cm artificial levels. Section G1 was closest to the sea (see Figure 11B). Excavation 2 yielded only a single stratum of brown clay accompanied by shells, predominantly nerites, along with pieces of charcoal, pottery, and stone artifacts until a depth of 60 to 80 cm below the surface. A gully, filled with refuse, could be observed to a depth of 140 cm along the G5–H5 line. According to Rouse (1947, fig. 1, 1953: 30, 69, 89, 103–104; Figure 14B), ceramics exclusively of the Cedros complex were recovered from Excavation 2. However, Cedros complex pottery seems to be restricted to the lower part of this trench, levels 5 to 7 (80 to 140 cm); its medium portion, levels 3 and 4 (40 to 80 cm), yielded ceramics transitional between the Cedros and Palo Seco complexes, whereas its upper part, levels 1 and 2 (0 to 40 cm), are characterized by Palo Seco pottery accompanied by a single Erin complex "trade" piece (Figures 47, 48, and 49). In contrast, the lower stratum of Excavation 1, levels 5 to 7 (80 to 140 cm), yielded Palo Seco complex pottery, whereas Palo Seco ceramics, accompanied by Erin "trade" pieces next to a single sherd of the Arauquinoid St. Catherine's complex, characterize its upper stratum, levels 5 to 7 (80 to 140 cm), all in addition to bone and stone artifacts (Figures 46, 68C–E, G, and 69A). Rouse's finds from the two excavations that included slightly more than 6,000 artifacts, almost evenly divided between them (Faber-Morse 2010), are in the Yale Peabody Museum collections (YPM ANT 174153–177009 and ANT 177010–180143). The zoological remains he recovered at Palo Seco were examined by Wing (1962:39–40, 46, 48–51; see also Newson 1976:48, 51–54, 66). Wing visited the site in July and August 1959, collecting pottery and ceramic artifacts from the surface that are now in the Florida State Museum.

Rouse, Olsen, and Cruxent visited the Palo Seco site in September 1969 to obtain samples for radiocarbon dating. They excavated three test pits in Midden 1, north and east of the camp school, where Donax shells appeared on the surface. As these pits yielded only thin shell deposits, a 2 by 1 m excavation unit (Pit D4), controlled in 25 cm artificial levels, was dug at a spot farther north of the school, somewhat lower down the gentle slope toward the road. Here they encountered refuse to a depth of 125 cm below the present surface. Cedros complex pottery was reportedly found in the bottom level (100 to 125 cm depth), capped by Palo Seco complex ceramics (Olsen 1969b, 1973, 1974:247, 253; Faber-Morse 2010). In addition, part of a human burial accompanied by a decorated vessel and some turtle carapace fragments were found in the 75 to 100 cm level (Olsen 1969b; not in the top level as reported by Olsen 1973). Pieces of charcoal "about the size of a grain of wheat or smaller" were collected from each level. A sample of charcoal recovered from the bottom level (100 to 125 cm depth), associated with Cedros complex ceramics, returned a radiocarbon date of 2060±80 BP (IVIC-641), or 176 cal BC–

cal AD 22. Furthermore, charcoal samples collected from the 75 to 100 cm, 50 to 75 cm, and 75 to 50 cm levels, associated with Palo Seco complex pottery, yielded radiocarbon dates of 1990±70 BP (IVIC-640), or 91 cal BC–cal AD 84, 1480±70 (IVIC-639), or cal AD 469–650, and 2130±80 BP (IVIC-638), or 351–49 cal BC. The third result was unexpected, because it was older than the results of the three samples collected from greater depths. According to Rouse (pers. comm. to Douglas Archibald 1974), this early date may be caused by the digging for a "burial which was found nearby," which brought up charcoal from the bottom of the site to its top levels (see Olsen 1973, 1974:235–236, 242–243, 247–253; Boomert 2000:519). Unfortunately, the archaeological specimens collected from Pit D4 were lost during shipment to Venezuela (TTHS Annual Report 1969; Archibald 1971:20; Boomert 2000:10). Finally, the situation of the Palo Seco site was inspected during a visit in March 1993 by Saunders and Chauharjasingh (2003:54–55), followed by Harris and Boomert in August 2004.

Note that the surface collection of pottery made at Palo Seco by Wing in 1959 (Wing 1962:39) consists entirely of ceramics of the Bontour complex. This material, which is curated in the Florida State Museum, includes in all 278 potsherds and 5 pottery artifacts marked "TPD Guesthouse." They may derive from a not yet relocated late-prehistoric midden deposit at the Palo Seco site. However, as Bullbrook and Rouse encountered Bontour potsherds at this site only in association with Erin and Palo Seco materials, it is possible that Wing collected this material at the Palo Seco East (SPA-8) site, a midden deposit some 1.2 km, as the crow flies, to the southeast of the Palo Seco settlement site (see Figure 11A). The Palo Seco East site was discovered by Bullbrook while he was excavating at Palo Seco in 1919. It was later investigated by Louis Winkler around 1950. The site's present situation was inspected during a visit in March 1993 by Saunders and Chauharjasingh (2003:38). Palo Seco East occupies part of a mud cliff on the beach of Palo Seco Bay, about 600 m east of the far end of Palo Seco Beach Drive. The site is continually being washed by the sea and indeed most of it has been eroded. It reportedly yielded predominantly Donax clams (*Donax* spp.) and Trigonal tivelas (*Tivela mactroides*), along with pottery of the Bontour complex (Bullbrook 1953:29, 65, 93, 1961a:5, 18; Rouse 1953:104).

Quinam

The multicomponent Ceramic settlement site of Quinam (SPA-10) is on the shore of the Columbus Channel, some 300 to 350 m west of Palmiste Point (Figure 15). Quinam (also called Palmiste) was discovered by Georges P. Wall and James G. Sawkins during their geological survey of Trinidad in 1858. However, they did not recognize it as an Amerindian shell midden and interpreted the site as a "raised beach," that is, a natural phenomenon (Wall and Sawkins 1860:66–67, fig. 5). It was rediscovered by T. C. Higgins, a geologist, about 1940 (Rouse 1953:104). The site consists of at least three separate shell refuse deposits, predominantly composed of Trigonal tivelas (*Tivela mactroides*), less of Donax clams (*Donax* spp.).

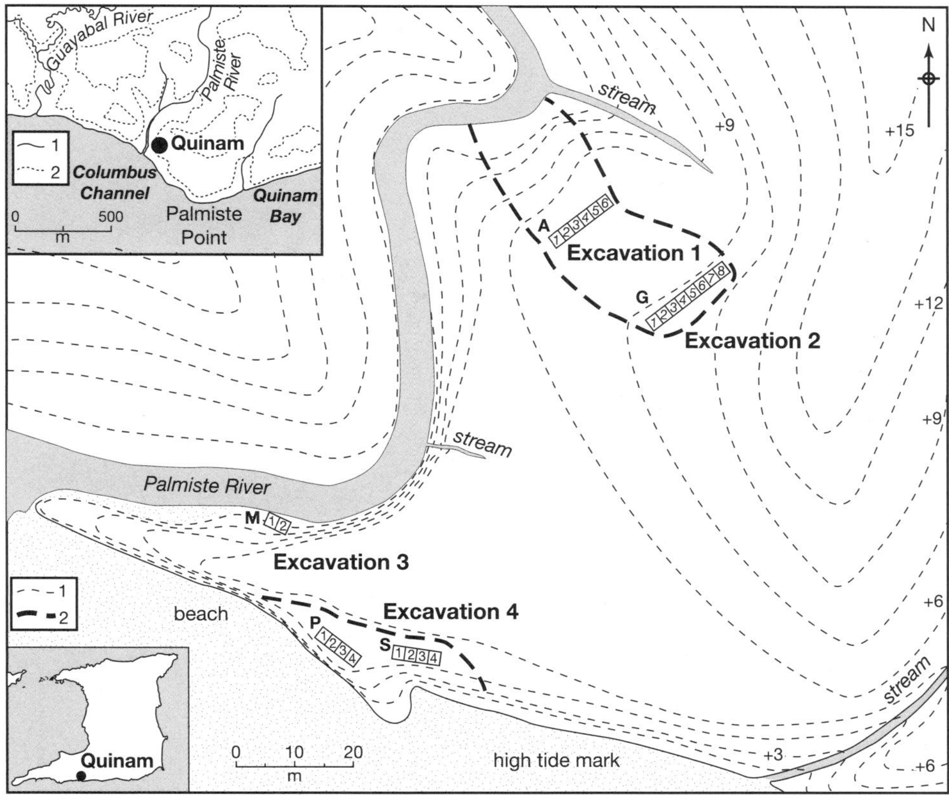

Figure 15. The Quinam (SPA-10) shell midden site in southwest Trinidad, showing Rouse's 1946 Excavations 1, 2, 3, and 4. *Legend:* (1) local contours; (2) boundary of shell midden refuse. Adapted from Rouse Papers, Quinam field notes. *Inset:* The geography of the Quinam archaeological site. *Legend:* (1) streams; (2) 25- and 50-foot contours.

The largest of these deposits (Midden 1), estimated to measure some 15 by 50 m, occupies the northwestern slope of a 5 to 10 m high, more or less triangular, plateau forming part of a ridge situated just east of the lower reaches of the Palmiste River, about 200 m inland from its mouth. This maximally 50 m high, more or less northeast–southwest ridge in Trinidad's Southern Range reaches the coast at Palmiste Point. It is covered with semi-evergreen seasonal forest. At the location of Midden 1 the slope is some 30° to 40°, and farther south the land falls more sharply to the river. Midden 2, estimated to measure some 6 m in diameter, is at the mouth of the Palmiste River, some 35 to 40 m to the southwest, whereas Midden 3 occupies the sandy beach, 6 to 20 m to the east of the second deposit. A seasonal stream, which debouches into the Palmiste River, forms the northern boundary of Midden 1, while another stream flows into the sea at the southeastern end of the plateau (Rouse Papers, Quinam field notes). The coast is eroding here. At high tide the sea reaches the base of the site at Midden 3; at low tide the water is 15 to 30 m away from the bluff, uncovering sand and a few boulders. The Palmiste River, which is brackish at its mouth, is cut off from the sea by a sand spit in the dry season. It does

FIGURE 16. View in 2004 of Trinidad's south coast near the Quinam archaeological site.

not flow into the Columbus Channel except after heavy rains. The sandy beach in front of the site offers a good landing place (Figure 16). It was occasionally used by Warao Indians from the Orinoco Delta in Venezuela, who until the 1940s came to Trinidad on trade expeditions and pilgrimages to Naparima Hill (Boomert 2000:21, 27–28, 424). (Rafts with escaped prisoners from the penal colonies in French Guiana used to drift with the South Equatorial Current from Devil's Island to Trinidad's south coast, making landfall on the beaches of Quinam, Palo Seco, Erin, or Icacos. In fact, during Rouse's work at Quinam in 1946, two escapees landed there [John A. Bullbrook, pers. comm. to Rouse].)

In the spring of 1942, Major (then Captain) J. E. L. Carter and other members of the Archaeological Section of the Trinidad and Tobago Historical Society, including Bullbrook and Kenneth W. Barr, started to excavate Midden 2 on the east bank of the Palmiste River, below the bluff. At the time the site was owned by Trinidad Leaseholds, Ltd. Carter's trench, which measured some 6 by 3 m, extended directly up to the river. Encountering shells, animal bones, and pottery of the Palo Seco complex, Carter abandoned digging before reaching sterile soil. His finds are in the National Museum of Trinidad and Tobago (the former Royal Victoria Institute and Museum) and a sample collection of potsherds was donated by Bullbrook to the University Museum of Archaeology and Anthropology at the University of Cambridge in the United Kingdom.

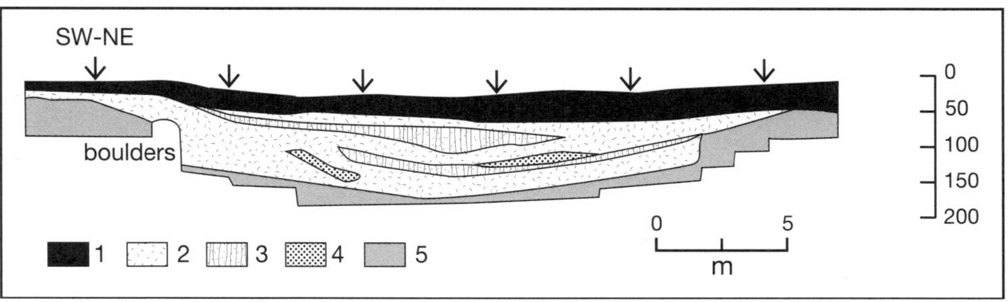

Figure 17. Stratigraphic profile of Rouse's Quinam Excavation 1 (for explanation, see text). *Legend:* (1) black humus; (2) medium brown clay, moderate shells; (3) shell concentrations; (4) light brown clay, rare shells; (5) light brown clay, sterile. Depth is in centimeters. Adapted from Rouse Papers, Quinam field notes.

Rouse later investigated the Quinam site. In July–August 1946 he first made a series of auger holes across the plateau to identify the densest refuse deposits and afterward dug five trenches, in 20 cm artificial levels, at the site. Rouse's Excavations 1 and 2 cut through the entire deposit of Midden 1. Excavation 1, measuring 12 by 2 m, was dug along the edge of the plateau. It consisted of six 2 by 2 m units (Sections A1 to A6) with a thick layer of dark brown to black humus on top of a deposit of medium brown clay containing many shells, with animal bones, pieces of charcoal, and Amerindian pottery, stone, bone, and pottery artifacts, resting on sterile light brown clay which, in Sections A3 and A4, extended to 160 cm below the modern surface. The vertical cross section showed the contours of a gully, falling from either end of the trench to a deep spot in the center, which had been filled in entirely with debris (Figure 17). A perceptible stream of clear water was flowing from the base of Section A4, which had many waterworn stones at its bottom. Finally, a group of isolated, disarticulated human bones, probably a secondary (partial) burial, was encountered at a depth of 20 to 40 cm in Section A2. Excavation 2, measuring 16 by 2 m and consisting of eight 2 by 2 m units (Sections G1 to G8), was dug at the upper side of Midden 1, parallel to Excavation 1. It showed the same stratigraphy as Excavation 1, with a shell deposit reaching down to 120 cm below the modern surface (Rouse Papers, Quinam field notes). Both trenches yielded pottery of the Palo Seco and Erin complexes throughout the deposit (YPM ANT 180144–182432 and ANT 182433–185286), along with bone and stone artifacts (Figures 50 to 55, 68A, H–J, 69D–J; Rouse 1947, 1953:104). Pure Barrancoid refuse was recovered only from the 0 to 20 cm levels of Excavation 1, as well as some potsherds that may represent "trade" materials of the St. Catherine's complex of the Guayabitan subseries, Arauquinoid series.

In the southwestern part of the Quinam site, Rouse dug an extension to Carter's excavation of Midden 2. This Excavation 3 consisted of a trench parallel to the Palmiste River, measuring 4 by 2 m. It was divided into two 2 by 2 m units (Sections M1 to M2) controlled in 20 cm artificial levels and reached sterile, light brown sand at a depth of 140 cm below the modern surface. The deposit appeared to be composed of medium brown sandy soil with different amounts of shell. Blue clay

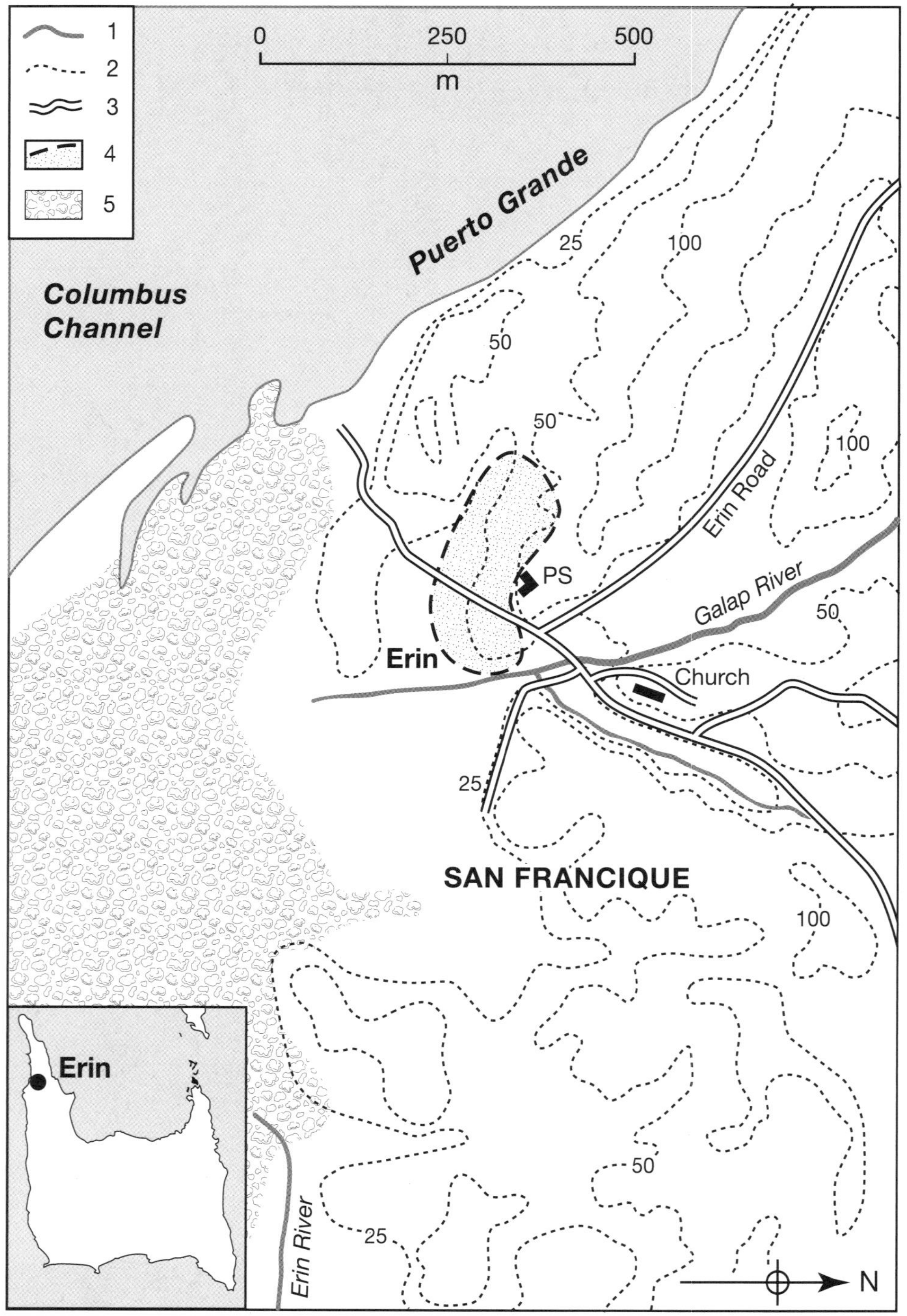

Figure 18. The geography of the Erin (SPA-20) archaeological site in southwest Trinidad. *Legend:* (1) rivers; (2) 25-, 50-, and 100-foot contours; (3) modern roads; (4) boundary of shell midden refuse; (5) mangrove woodland. PS, police station. Partially adapted from Rouse Papers, Erin field notes.

or rocks at the bottom of the refuse had apparently washed downhill. Excavation 3 yielded ceramics along with stone and pottery artifacts of the Palo Seco complex (YPM ANT 185287–185758), as well as animal bones and flint chips (Figure 56; Rouse 1947, Papers, Quinam field notes). Finally, Rouse investigated Midden 3 on Quinam beach by digging two separate, more or less northwest–southeast trenches, each measuring 8 by 2 m (Excavation 4). Both trenches were divided into four 2 by 2 m units (Sections P1 to P4 and S1 to S4) controlled in 20 cm artificial levels. Excavation 4 yielded a deposit of dark brown humus turning into sterile, light brown clay resting on boulders at a depth of 20 to 30 cm. The refuse contained few shells, animal bone fragments, and pieces of charcoal, as well as 493 potsherds next to pottery artifacts and stone chips (YPM ANT 185759–185947, ANT 256027–256030) belonging to the Bontour complex (Figure 67F–H; Rouse 1947, Papers, Quinam field notes; Boomert 1985). All the materials excavated by Rouse at the Quinam site are in the Yale Peabody Museum. His zoological finds were analyzed by Wing (1962:42–43, 46, 1977; also Newson 1976:48, 51–54; Boomert 2000:339–340). In addition to Rouse's work at Quinam, in December 1967, January 1968, September–November 1969, and February 1970 members of the Trinidad and Tobago Historical Society (South Section), notably Harris, inspected the site, collecting surface materials and making experimental digs (TTHS Annual Report 1969; TTHS Newsletter 1970(2), 1973(2); Harris 1972:11–13). Their finds, including shells, animal bone fragments, Palo Seco pottery (in all 104 potsherds), and stone artifacts, are at the University of the West Indies. In March 1993, Saunders and Chauharjasingh (2003:39–40) inspected the site, followed by Boomert and Harris in August 2004. The latter noticed that all of Rouse's excavations can still be identified; apparently no attempt was ever made to fill them. We can conclude that the Ceramic Amerindians occupied the center of the Quinam plateau, discarding their refuse on its west, north, and south edges. Geologically, the site is characterized by the clays and silts of the Cruse Formation, dating from Miocene times. Soils are loamy, dominated by typic haplustults (ultisols) of the Moruga Clay variety. Precipitation amounts to 125 to 150 cm per year.

Erin

The Ceramic settlement site of Erin (SPA-20) represents a multicomponent shell midden deposit, measuring at least 260 by 142 m, next to and behind the police station in the village of San Francique, roughly 10 km west of Palo Seco. The site is about 400 m inland from the beach of Puerto Grande on the Columbus Channel and some 500 m northwest of the mouth of the Erin River and its adjacent mangrove tidal woodland (Figure 18). It occupies the northern plateau and the wide, seaward flank of a ridge about 25 m high, belonging to Trinidad's Southern Range, which encircles part of the alluvial flat along the left bank of the lower Erin River (Figure 19). This northwest–southeast ridge runs parallel to the shore of the Columbus Channel. A seasonal right tributary of the Erin River, locally known as the

FIGURE 19. View in 2004 of the Erin archaeological site in southwest Trinidad.

Galap River, flows north of the site. It is the closest source of potable water to the site, since the lower reaches of the Erin River are brackish. The deposit, predominantly consisting of Trigonal tivelas (*Tivela mactroides*) and Donax clams (*Donax* spp.), is most dense at the hill foot where it approaches 1.5 m in thickness (Boomert 2000:147–148, 203, 357). It rests predominantly on Late Miocene–Early Pliocene sandstone beds belonging to the Morne l'Enfer Formation. The area is typified by intermediate upland soils, notably udorthentic chromusterts (vertisols) and typic haplustults (ultisols) of the La Retraite and Moruga Clay varieties, respectively. To the north of the Erin site these soils are replaced by deep alluvial soils with restricted internal drainage, notably entic pelluderts (vertisols) of Cromarty Clay type. Originally, the scenery of the site was dominated by semi-evergreen seasonal forest. Today part of this is under cocoa cultivation, used for household gardens, or built over. According to Bullbrook (1953:93), there is some evidence that the shore is rising here. Precipitation amounts to 125 to 150 cm per year. The site forms the type site of the Erin complex belonging to the Erinan subseries of the Barrancoid series.

The Erin site was discovered in May 1888 by Sir William Robinson, the then governor of Trinidad, and H. Fowler, the colonial secretary, who noticed potsherds and shells in a road cutting at San Francique during an official visit to the village. Correctly judging these finds to be pre-Columbian, they instructed the local warden, A. Newsam, to collect as many objects from the site as he could find (Bullbrook 1960:6–7). The attractively decorated pottery picked up from the surface by Newsam and others sparked much interest and ultimately found their way to the newly established Royal Victoria Institute and Museum (now the National Museum of Trinidad and Tobago) in Port-of-Spain. Choice pieces were illustrated by Collens (1888) and Fewkes (1907:190–191). About the turn of the twentieth century, the Rev. Thomas Huckerby put together a substantial collection of individual finds from Erin, including pottery, stone, and shell artifacts, on behalf of the Heye Foundation, now the National Museum of the American Indian, in New York, apparently by buying specimens locally in San Francique (Fewkes 1914; "Guide to the Collections…" 1922:36–37; Anna C. Roosevelt, pers. comm. to Boomert 1984). Intrigued by Huckerby's finds, Jesse Walter Fewkes investigated the Erin site during a collecting trip to the Caribbean under the joint auspices of the Bureau of American Ethnology and the then Museum of the American Indian in the wet season of 1913. At the time the site was being dug for shells used for road gravelling by the colony's Public Works Department. Fewkes excavated at an unspecified location in the western part of the flat, locally known as the "savannah," at the base of the ridge, noting that the deposit showed shell layers alternating with vegetable mold, ashes, and humus (Fewkes 1914, 1922:62–78; Bullbrook 1953:82–83). Fewkes encountered pottery, stone, and bone artifacts of the Palo Seco and Erin complexes, now in the National Museum of the American Indian (Anna C. Roosevelt, pers. comm. to Boomert 1984). Additional potsherds were collected on behalf of the Museum of the American Indian by de Booy during a surface search of the site from May to September 1915 (de Booy 1917). Destruction of the site due to digging for shells by the Public Works Department finally stopped in 1916–1917.

After fire had destroyed the Royal Victoria Institute and Museum and its holdings in 1919, Bullbrook was asked to acquire archaeological specimens from the Erin site to form a new collection for the museum. In 1932 Bullbrook took up residence in Erin and started to investigate the site more thoroughly. Intermittently between April 1934 and July 1935 he excavated a trench of unspecified dimensions at the eastern end of the site, slightly west and behind the police station "as close to the buildings as possible," encountering black soil containing many shells, animal bones, and Erin as well as Palo Seco complex pottery (Bullbrook 1940:10–11, 1953:9; Rouse Papers, Erin field notes). Subsequently, the Archaeological Section of the Historical Society of Trinidad and Tobago received a small grant from the colonial government to continue the Erin investigations (Bullbrook 1941; Espinet 1950). Intermittently between April 1941 and October 1942 Carter and Barr dug an irregular trench, measuring about 12 m in length and 1.4 to 3.0 m in width, on the hillside above the flat, just below the plateau formed by the hilltop, on the opposite side of the plateau from the police station and Bullbrook's 1934–1935 excavation. While reportedly recognizing "two strata" in the midden (Osgood 1942), Carter and Barr recovered shells, animal bones, and ceramics along with stone and fossil coral artifacts of the Palo Seco and Erin complexes. They also collected a few pot-

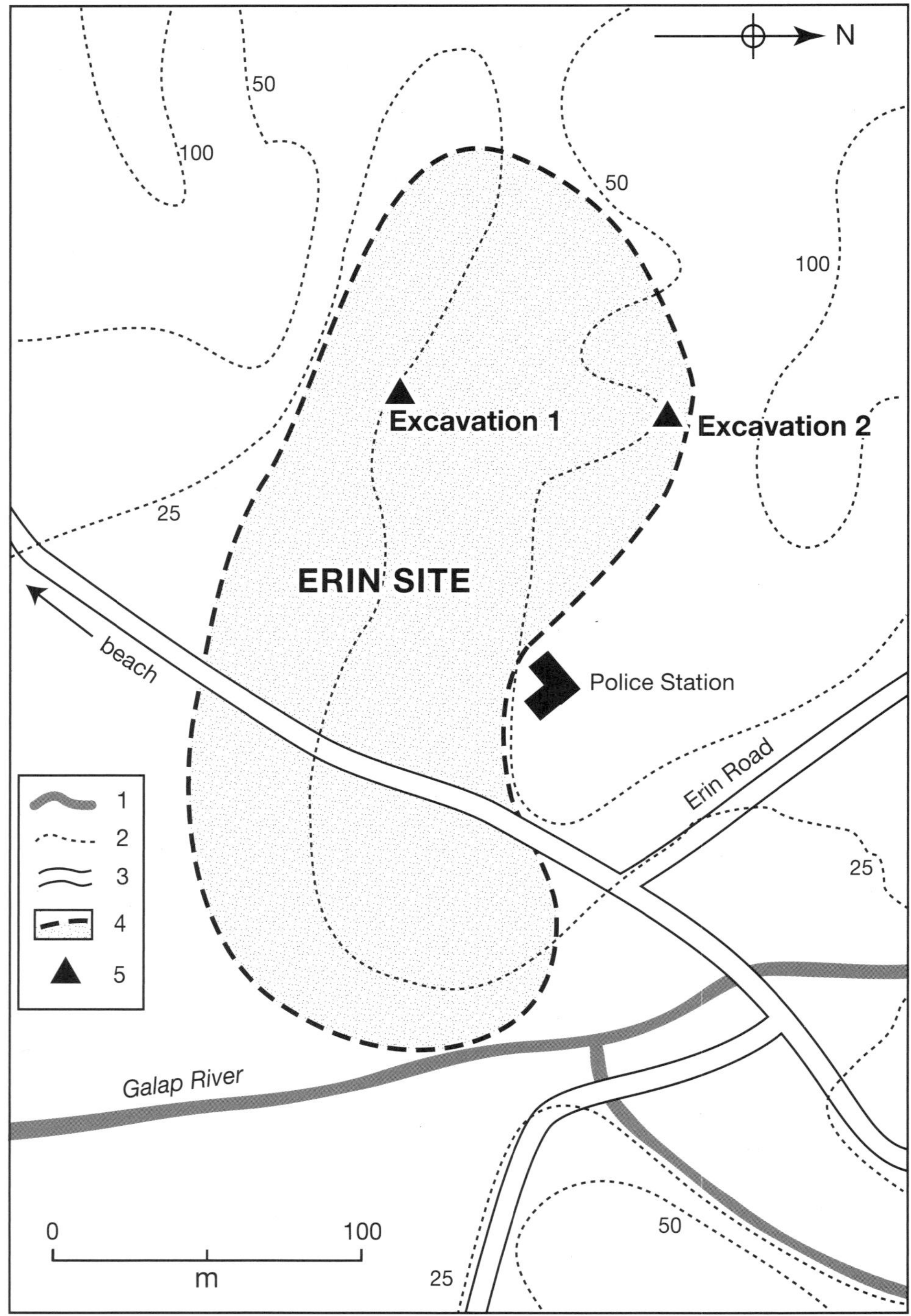

FIGURE 20. The Erin shell midden site in southwest Trinidad, showing Rouse's 1946 Excavations 1 and 2. *Legend:* (1) streams; (2) 25-, 50-, and 100-foot contours; (3) modern roads; (4) boundary of shell midden refuse; (5) site of excavation trenches. Partially adapted from Rouse Papers, Erin field notes.

FIGURE 21. Rouse's 1946 Excavation 1 in progress at the Erin site. Note the dense layers of shell visible in the vertical section of the trench. Photograph by Irving Rouse.

sherds that can be assigned to the St. Catherine's complex of the Arauquinoid series with Apostadero "trade" sherds. Most of these finds are in the National Museum of Trinidad and Tobago; a small collection was donated by Carter to the Yale Peabody Museum when he visited Cornelius Osgood in the spring of 1942.

Much charcoal was encountered, especially toward the bottom of the midden (Howard 1943:65; Bullbrook 1953:93). In July 1941, Carter and Bullbrook found a flexed primary inhumation of a young adult male, interred on his right side, at some 30 cm below the surface "high up the hill, near the apex of the midden material." It was accompanied by a large bowl with hollow rim (originally a rattle), a hammerstone, a small bowl with flanged rim, and a stone axe head. The entire burial was salvaged and has been on exhibition in the National Museum of Trinidad and Tobago ever since (Bullbrook n.d., 1949, 1960:8, 49; Feriz 1959:85–86; Bullen 1970; Boomert 2000:158, 399; Barr, pers. comm. to Peter O'Brien Harris).

The incentive for further investigations was brought about by the accidental find of the famous Erin "water bottle" by a villager who was digging for adobe (*tapia*) on the flat land south of the midden deposit in the central portion of the site in April 1944 (Sleight 1946; Espinet 1950; Bullbrook 1953:40–41, 1960:49, 1961a:14, 16, pers. comm. to Rouse; Rouse 1953:34). This asymmetrical ceremonial vessel, now in the National Museum of Trinidad and Tobago, originally belonged to the mortuary offerings, which also include three similar white-on-red painted bowls, of a probably male primary inhumation (Boomert 2000:159, 399). Its discovery led to the excavation by Bullbrook and Carter of an irregular, 60 to 90 cm wide trench between Bullbrook's old excavation ("almost touching it") and the beach road between February and July 1945. It showed dark soil, yielding few shells but many animal bones, as well as pottery, stone, bone, shell, and fossil coral artifacts of the Palo Seco and Erin complexes, until the bottom of the deposit was reached, about 45 cm below the present surface (Bullbrook 1953:53, 1961a:1–2). Most of these finds are in the National Museum of Trinidad and Tobago; a small collection was donated by Bullbrook to the University Museum of Archaeology and Anthropology at the University of Cambridge. A few Apostadero potsherds marked BE ("Bullbrook Erin") in the National Museum of Trinidad and Tobago collection, clearly representing "trade" pieces, may derive from this same excavation. The animal bone materials recovered during the Bullbrook–Carter excavations of 1945 were analyzed by Wing (1962:40–42). According to Bullbrook (1949, 1960:49–50), he encountered eight human burials at Erin. At least two of these were flexed interments accompanied by mortuary gifts, both probably belonging to the Palo Seco complex (Bullbrook n.d., 1953:40–41, 1960:49). He also notes that, apart from these apparently undisturbed primary burials, several secondary deposits of bundles of human bones and a single skull were found (Bullbrook n.d., 1940:39–40, 1949, 1953:50).

In July 1946, Rouse investigated the Erin site by excavating two trenches, controlled in 20 cm artificial levels (Figures 20 and 21). First, a T-shaped trench (Excavation 1) measuring 8 m by 2 to 6 m in the central portion of the flat, at the foot of the ridge was dug. It showed a large concentration of shells and was far enough from the hillside to avoid wash. Excavation 1 consisted of six 2 by 2 m units, of which four (Sections B1 to B4) stretched along the long axis of the trench, whereas two other units (A4 and C4) were joined at both sides of B4 to form its short axis (Figure 22). This trench yielded dark brown humus containing shells and archaeological materials, alternating with pure shell layers, resting on sterile yellowish brown clay at a depth of 150 cm below the surface. Charcoal was most frequent in the lower portion of the deposit.

Two badly decayed inhumation burials without associated mortuary gifts were encountered in the 80 to 100 cm level of Section C4, while a third, similarly lacking such gifts, was found in the 80 to 100 cm level of Section B3 (Rouse Papers, Erin field notes). Only one of these burials, a flexed inhumation, could be reconstructed. One of the Section C4 interments seemed to be deposited on his or her right side parallel to the long axis of the trench with the head bearing northwest. On

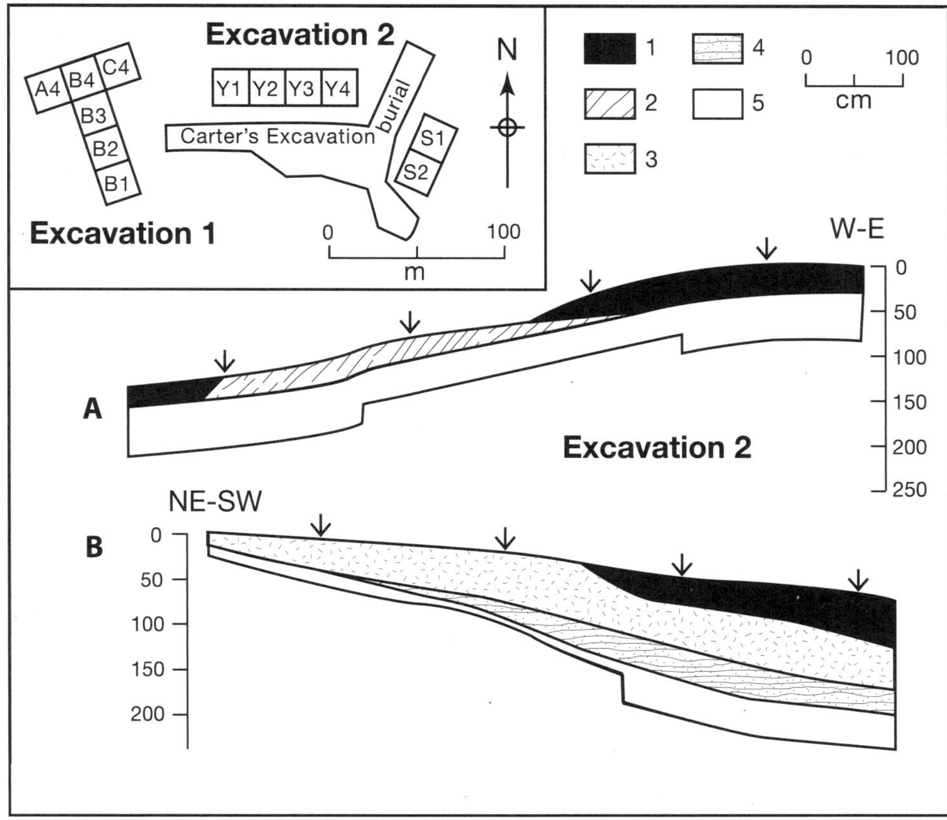

FIGURE 22. Stratigraphic profiles of Rouse's Erin Excavation 2 (for explanation, see text). **A,** Sections Y1 to Y4. **B,** Cross section at far end of Carter's excavation (next to Rouse's Sections S1 and S2). *Legend:* (1) dark brown clay, few shells; (2) dark brown clay, moderate shells; (3) light brown clay, few shells; (4) light brown clay, many shells; (5) light brown clay, sterile. Depth is in centimeters. *Inset:* Rouse's Erin Excavations 1 and 2. Adapted from Rouse Papers, Erin field notes.

the basis of their stratigraphic position, the burials can be dated to Palo Seco times. Excavation 1 yielded stone and pottery artifacts of the Palo Seco and Erin complexes (Rouse 1947, 1953) with some Bontour complex pieces and "trade" sherds of the St. Catherine's complex of the Arauquinoid series in its top levels (Figures 59, 60, 61, and 68B, F). Rouse's second trench (Excavation 2) was dug into the slope of the hill, "alongside" the Carter and Barr's trench of 1941–1942 (see Figure 22). Its two separate cuts, measuring 8 by 2 m (Sections Y1 to Y4) and 4 by 2 m (Sections S1 and S2), yielded dark brown humus containing shells and potsherds of the Palo Seco complex, with some Erin and Bontour complex pieces in the top levels (Figure 62). Sterile yellowish clay was reached at depths of 80 and 130 cm, respectively. A disturbed inhumation burial was encountered in the 0 to 20 cm level of Section Y1 (Rouse 1947, 1948, Papers, Erin field notes). All finds from Rouse's Excavations 1 and 2 at Erin are in the Yale Peabody Museum (YPM ANT 168783–173169 and ANT 173170–174152).

After Rouse's work at Erin, the site was only visited intermittently by interested researchers. For instance, at an unreported date, Father Robert Pinchon of Martinique made a *"sondage rapide"* at Erin, apparently encountering Palo Seco pottery, which he compared with the Saladoid ceramics of the Lesser Antilles (Pinchon 1952). According to Bullbrook (pers. comm. to Rouse), amateur investigators found Bontour complex pottery in the side "of our lower pit" at Erin in January 1952. In addition, a small surface collection of Erin and Bontour complex potsherds from the site, now at the Florida State Museum, was made by Wing during a visit in July 1959. Wing also analyzed the animal bone materials recovered by Rouse during his 1946 excavations (Wing 1962:40–42; see also Newson 1976:52–53; Boomert 2000:339–340). Members of the South Section of the Trinidad and Tobago Historical Society, notably Harris, Basil E. Josa, and Clyde Rampersad, made several visits to the Erin site in 1969, 1970, and 1972, excavating a series of five 1 by 1 m test pits that yielded dark clayey soil containing shells and animal bones with pottery and stone artifacts of the Palo Seco and Erin complexes (TTHS Annual Report 1969; TTHS Newsletter 1970(3, 4), 1972(4), 1973(3)). This collection is kept at the University of the West Indies. Several pottery finds were made by Carla Bridglal of La Romain during the construction of a new police station on part of the Erin site in the 1990s. Saunders and Chauharjasingh (2003:47) visited the site in May 1997, collecting potsherds and a fragment of human bone from the surface of the site. Finally, the current condition of the site was inspected by Boomert and Harris in August 2004.

Bontour

The Ceramic settlement site of Bontour (VIC-13) was situated on a plateau about 25 to 30 m high behind the present General Hospital buildings in San Fernando, some 150 m east of Bontour Point on the shore of the Gulf of Paria (Rouse 1953:100). This plateau is abruptly cut off at the seaside, with a steep cliff on its western edge that stretches along the present Sea Wall and Lady Hailes Avenue (Figure 23). Bontour is the type site of the late-prehistoric Bontour complex, belonging to the Guayabitan subseries of the Arauquinoid series. It consisted of an extended, but relatively thin, deposit of shells covering much of the plateau, which in the 1940s was partially used as a household garden. At the time the Bontour plateau was owned by the now defunct Trinidad Government Railways, a major track of which ran parallel to the shore of the Gulf of Paria at the base of the bluff. The plateau shows a slight slope toward the southern end of the site; here the shell deposit, predominantly consisting of blood arks (*Anadara ovalis*) and Caribbean oysters (*Crassostrea rhizophorae*), seemed to be thickest (Rouse Papers, Bontour field notes). A fault breccia once existed beneath the cliff at Point Bontour, with an outcropping of flint and chert (John A. Bullbrook, pers. comm. to Rouse).

In 1953, most of the site was bulldozed during construction of the hospital and there was further destruction when new hospital structures were built in the

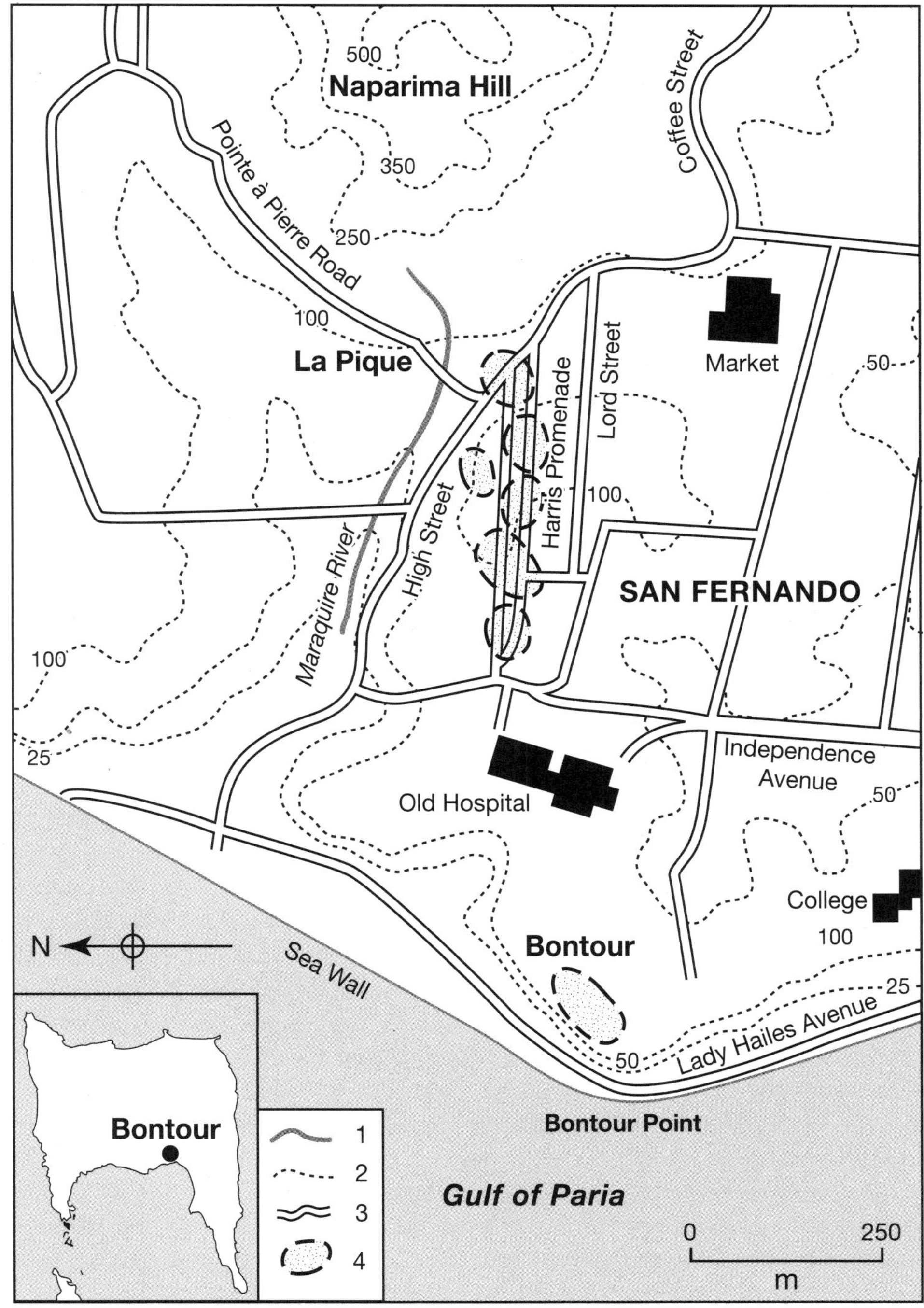

Figure 23. The geography of the Bontour (VIC-13) and Harris Promenade/High Street (VIC-14) archaeological sites, San Fernando, in southwest Trinidad. *Legend:* (1) rivers; (2) 25-, 50-, 100-, 250-, 350-, and 500-foot contours; (3) modern roads; (4) shell midden refuse. Partially adapted from Rouse Papers, Bontour field notes and Chauharjasingh, pers. comm. 2005.

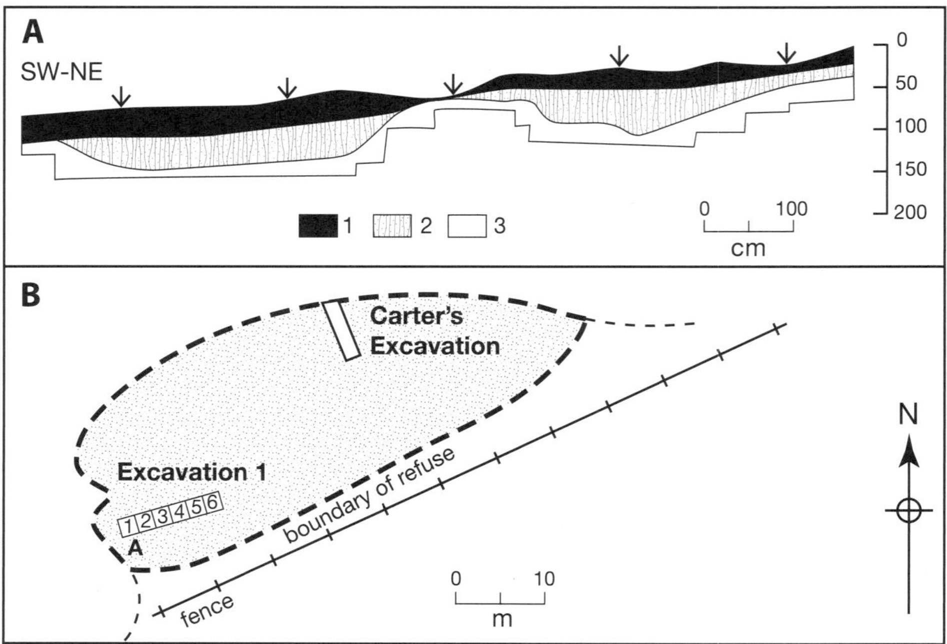

Figure 24. The Bontour site. **A,** Stratigraphic profile of Rouse's Bontour Excavation 1 (for explanation, see text). *Legend:* (1) black humus, few shells; (2) humus, moderate shells; (3) yellowish clay, sterile. Depth is in centimeters. **B,** Rouse's Bontour Excavation 1 and Carter's excavation. Adapted from Rouse Papers, Bontour field notes.

1980s. The entire site has been destroyed, the last remaining areas covered by asphalt for use as a car park in 1992. The nearest source of potable water is a ravine some 500 m to the north of the site. This formerly carried a semi-permanent stream, locally known as the Maraquire River, which flowed from Naparima Hill (178 m) to Bontour Bay (Wise 1934–1938, 1:30). The present High Street, in the center of San Fernando, follows the southern bank of this ravine (the street name is significant, of course). Geologically, the Bontour area is characterized by the calcareous clays and marls of the Cipero Formation dating from Oligocene to Miocene times. Soils are very fine-grained and montmorillonitic, dominated by aquentic chromuderts (vertisols) of the Princes Town Clay variety. Originally this part of Trinidad's Southern Lowlands supported deciduous to semi-evergreen and evergreen seasonal forest. Local rainfall amounts to 150 to 175 cm per year.

The Bontour site was known as early as 1934 (Wise 1934–1938, 1:30). In March 1942, it was investigated by Carter, assisted by Barr and other members of the Archaeological Section of the Historical Society of Trinidad and Tobago. Carter dug a northwest–southeast trench measuring 3.60 by 0.90 m through the Bontour shell midden to sterile soil, encountering a deposit of reportedly some 20 to 30 cm depth containing Amerindian pottery (2,135 potsherds) and stone artifacts, with some modern ceramics and glass in its top levels (Rouse 1947, Papers, Bontour field notes). Carter subsequently reported on his finds to Osgood and Rouse. In August

FIGURE 25. Vertical section of Rouse's 1946 Excavation 1 at the Bontour site, showing dense shell layers resting on sterile yellowish clay. Photograph by Irving Rouse.

1946, Rouse dug a southwest–northeast 12 by 2 m trench (Excavation 1) controlled in 20 cm artificial levels in the southeastern part of the Bontour plateau, about 22 m to the southwest of Carter's excavation. This trench was divided into six 2 by 2 m units (A1 to A6) and extended inward perpendicular to the edge of the plateau (Figure 24B). Rouse encountered a layer of humus underlaid by a single, irregular semi-circular shell deposit, the long axis of which stretched along the top of the bluff. It consisted of black humus with scattered mollusks, increasing in frequency from north to south, yielding pottery of the Bontour complex (1,346 potsherds) along with stone, bone, and pottery artifacts, and scattered pieces of charcoal, as well as flint and chert chips (Figure 66). Most shells were recovered from the southwest part of the trench. At a depth ranging between 30 and 80 cm below the surface the shell deposit gave way to sterile, yellowish (marly) clay (Figures 24A and 25). The top level seemed to have been disturbed by plowing and yielded some

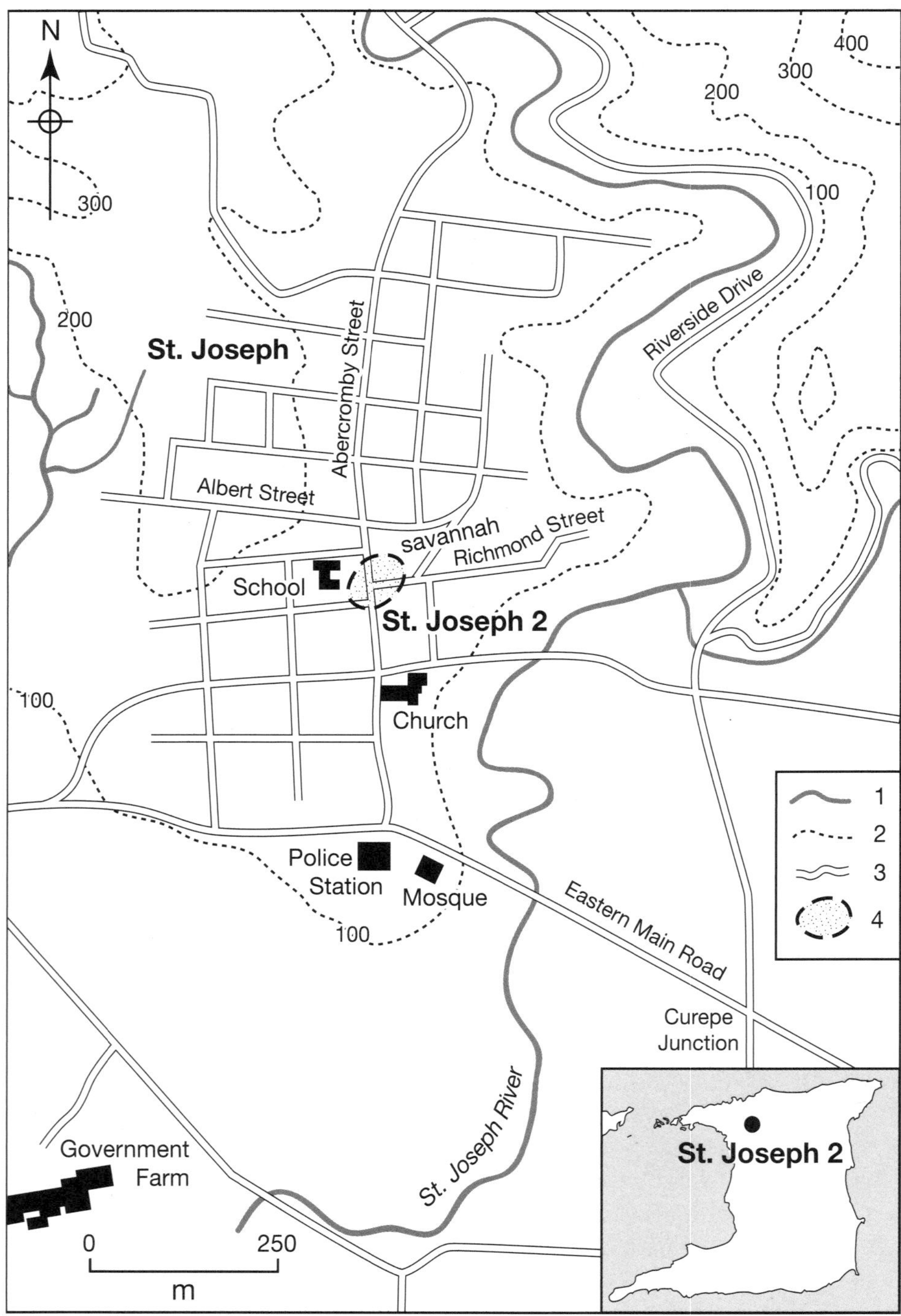

FIGURE 26. The geography of the St. Joseph 2 (SGE-16) archaeological site in northwest Trinidad. *Legend:* (1) rivers; (2) 100-, 200-, 300-, and 400-foot contours; (3) modern roads; (4) shell midden refuse. Partially adapted from Rouse Papers, St. Joseph field notes.

nineteenth-century European ceramics and glass (Figure 80; Rouse 1953:100–101, Papers, Bontour field notes; Appendix C).

Together with Bullbrook, Rouse returned to the Bontour site in August 1953 and collected a few more potsherds. After the bulldozing of most of the plateau for the construction of the new hospital in May–June and December 1981, surface finds of pottery and stone artifacts were salvaged from the site by Harris, Cecil Williams, and Boomert. The Bontour site was inspected by Chauharjasingh and Saunders in July 1992. By now the remaining parts of the plateau were being prepared for the construction of a car park. They returned to the site in February 1997, noting that most of it had been covered by new hospital buildings (Saunders and Chauharjasingh 2003:8–9; Chauharjasingh pers. comm. to Boomert 2005). The Bontour finds of both Carter and Rouse are curated in the Yale Peabody Museum (YPM ANT 163099–165090 and ANT 165091–166199), those of Harris and Boomert at the University of the West Indies. There are also a few pieces of pottery from the site in the collection of the National Museum of Trinidad and Tobago in Port-of-Spain.

Interestingly, some 450 m to the east of the Bontour site similar archaeological materials have been encountered during street excavations at Harris Promenade, extending to High Street, in the center of San Fernando. The extended shell midden deposit of this Harris Promenade/High Street site (VIC-14) stretches over a distance of some 400 to 450 m, from the police headquarters at the western end of Harris Promenade to as far east as its junction with High Street and Coffee Street (Figure 23). In 1970 and 1971 street excavations revealed a 10 to 30 cm thick shell deposit here, below some 50 to 60 cm of recent fill and resting on sterile, grayish clay. It was inspected by members of the Trinidad and Tobago Historical Society (South Section), notably Harris and Williams, who collected shells, predominantly Caribbean oysters (*Crassostrea rhizophorae*), Tiger lucinas (*Codakia orbicularis*), and blood arks (*Anadara ovalis*), along with animal bones, Bontour complex pottery, chert chips, and pieces of charcoal (Peter O'Brien Harris, pers. comm. to Boomert 1982; TTHS Annual Report 1971, 1976–1977; TTHS Newsletter 1970(3, 4), 1971(3, 8), 1976(4)). In October 1983, additional archaeological materials were salvaged from this site by Boomert. In May 1997 Saunders and Chauharjasingh visited the site, noting that there was ongoing construction associated with the building of a new police station. Construction workers informed them that Amerindian pottery had been found beneath the floor of the old station (Saunders and Chauharjasingh 2003:9). Chauharjasingh (pers. comm. to Boomert 2005) returned to the site in January 1998, inspecting its situation. All finds collected by Boomert, Saunders, and Chauharjasingh are in the University of the West Indies collection.

St. Joseph 2

The Ceramic–Historic settlement site of St. Joseph 2 (SGE-16), in the center of the modern town of St. Joseph in the northwestern part of Trinidad (Figure 26), consists of a shell midden deposit that occupies an area of approximately 50 by 50 m at both sides of Abercromby Street, between Richmond Street and Albert Street, including the western side of the present town "savannah" to the east of Abercromby Street and the easternmost part of the government schoolyard to its west. In the 1950s, shell refuse was present also in gardens of houses on Albert Street, north of the "savannah" (Rouse Papers, St. Joseph field notes). The site is located on a steep slope and small plateau, about 45 m above MSL, some 250 m west of the St. Joseph (Maracas) River, a right tributary of the Caroni River. It is actually on the first high rise north of the Caroni River valley, and as such the site is positioned at the boundary between Trinidad's Northern Range and the Northern Basin, offering a free view across the latter as far south as Trinidad's Central Range (Boomert 1984b:34). The town of St. Joseph occupies the entrance of the valley of the St. Joseph (Maracas) River, one of the several north–south flowing streams that drain Trinidad's Northern Range. Geologically, the site's area consists of the Pleistocene river terraces of the Cedros Formation, which become wider and more extended toward the south, in the Caroni River valley. They are flanked by the Cretaceous deposits that typify the hills and mountains of this portion of the Northern Range. Soils are kaolinitic and freely drained; they are considered to belong to the orthoxic tropudults (ultisols) of the St. Augustine group of clay loams. Precipitation varies between 150 and 175 cm per year. The natural vegetation of the St. Joseph region is characterized by semi-evergreen seasonal forest, toward the north grading into lower montane forest. In colonial times the flat, grassy plain to the south of St. Joseph was either used to grow sugar cane or left uncultivated, whereas the hilly slopes of Maracas Valley north of the town are traditionally under cocoa, coffee, and citrus cultivation, as well as used for household gardens.

St. Joseph was founded as San José de Oruña by Domingo de Vera in 1592 and functioned as Trinidad's Spanish capital until 1784, when it was replaced by Port-of-Spain. At the time of the Spanish occupation, a *cacique* of the Carinepagoto Indians, most likely a branch of the Kalina (Carib), called Goanagoanare (Guanaguanare, Wannawanare), "granted an area on the highlands where the Spanish could settle" and afterward "with his men withdrew to another part of the island," probably the San Juan region (Notarial Record 1592; Salazar 1595; Liaño 1596; Boomert 1986, 2002). The layout of the town would have followed the Spanish laws regarding new towns, which stipulated the location of a main square (*plaza*) in the town's center and the reservation of plots for the church, governor's palace, town hall (*casa de cabildo*), and prison. According to regulations proclaimed in 1573, in inland towns the church was not to be situated on the town square, but at some distance from it. This conforms with the location today of the Roman Catholic church in St. Joseph, built in 1815, which is one block to the south of the town "savannah." A tombstone in the churchyard that carries an inscription with the year 1682 indicates that this has definitely been the site of the church since the late seventeenth century (Joseph 1838:140; Newson 1976:117, 287). However, it is uncertain whether the present town "savannah" formed the *plaza* in Spanish times. Indeed, Wise suggested (1934–1938, 3:38, 4:39–40) that the original town square was to the west of the present church. It would have been formed by a rectangular area that gradually built up after the British conquest of Trinidad in 1797. Indeed, the open character of the present town "savannah" may have originated from its use as a military cemetery in the early nineteenth century. The moss-grown and broken tombstones of this graveyard were still to be seen in the first decades of the twentieth century (Telles 1927).

Throughout Spanish colonial times St. Joseph remained a small and impoverished town. It was sacked several times by Spain's enemies, starting with Sir Walter Raleigh in 1595. The town was connected with Port-of-Spain and the Aricagua (San Juan) *encomienda* to the west and the three *encomiendas* to the east—Cuara, Tacarigua, and Aruaca (Arouca)—by a bridle path that ultimately became today's Eastern Main Road. Its most western extension is still known as the Old St. Joseph Road. Only the northwestern part of Trinidad was effectively controlled by the small population of white and *mestizo* Spanish settlers on the island, who until the 1770s were outnumbered by the surviving groups of Amerindians who lived scattered throughout the rest of Trinidad. Plantations were few and the production and trading of cash crops such as tobacco and cocoa flourished only temporarily.

The Spanish capital reflected the impoverished state of the island. All buildings in St. Joseph were made of wattle-and-daub. A Dutch report of 1637 describes the town as consisting of only thirty houses "made of earth stamped solid which they call *tapias* and roofed with thatch or other combustible material" (Ousiel 1637). In 1681 the parish priest complained that the church was "made of *tapias* covered with palms and all so rough and unadorned and indecent for the celebration of divine worship and moreover it is old and a threatened ruin" (cited in Newsom 1976:117). Population remained small, reportedly varying from about thirty-five to fifty men in 1609 and 1637 to ninety to one hundred householders (apart from some tens of Amerindian servants and black slaves) in 1688 (Newson 1976:117, 121–122). After the outbreak of smallpox in 1739, St. Joseph was almost totally abandoned when the *vecinos* (householders with municipal rights) went to live in the countryside. In 1757 the newly arrived governor of the island called the town a derelict collection of mud huts, most of which were collapsed, and went to live in Port-of-Spain (Borde 1982:127–128). In 1766 an earthquake destroyed what was left of the town. It was only in the late eighteenth century that St. Joseph gradually became populated again: in 1797 it had 728 inhabitants. In 1837 the town was the scene of an uprising of the First West India Regiment, which was stationed there until it was disbanded in 1927.

The St. Joseph 2 site was discovered by Goggin in August 1953 and excavated by Rouse and the Goggins that same month. It yielded prehistoric Amerindian finds such as pottery, stone and bone artifacts, and a human burial of the Bontour complex, Guayabitan subseries, Arauquinoid series (Figures 67 and 69B, C), as well as historic period Amerindian ceramics of the Mayo complex, Mayoid series (defined by Boomert [1985]), and European materials (Figures 76–79, 83, 85, 86). The latter include Spanish, German, Dutch(?), English, and Chinese ceramics and other artifacts dating to the late sixteenth through nineteenth centuries (see Appendices B and C). Goggin and Rouse dug two 2 by 6 m trenches in 10 cm artificial levels at the site, each divided into three 2 by 2 m excavation units (Units 1A1-3 and 2A1-3). Trench 1 was in the schoolyard opposite the southwest corner of the town "savannah." It consisted of humus-stained dark brown clay mixed with small quartz and schist pebbles and archaeological materials down to a depth of about 50 cm. The occupation layer rested on yellow to orange sterile clay (Goggin Papers, St. Joseph field notes; Rouse 1953:97–98, Papers, St. Joseph field notes). Trench 1 yielded mostly European pottery and some Amerindian ceramics (315 potsherds). The European materials (some 2,500 sherds) include fragments of Spanish majolica and Middle Style olive jars dating from the late sixteenth through eighteenth centuries, along with delftware, chinaware, Rhenish stoneware, various nineteenth-century English ceramic styles, pipe fragments, and pieces of glass, including remnants of many gin and rum bottles (Silver 2009; Silver and Faber-Morse 2010; see Appendices B and C). In addition, several gunflints, bricks, a glass bead, some slate-pencils, hand-forged iron nails of "rosehead" type, knives, and fragments of an iron rake, hinge, padlock, and a horseshoe were found, as well as some harness ornaments made of brass, all dating from the nineteenth century (Appendix C). The site of Trench 1 is reported to have been the location of the former British military barracks at St. Joseph, explaining the abundance of nineteenth-century finds, including the most remarkable piece recovered from this trench, a star-shaped insignia marked "Trinidad Rifle Brigade" (Section A2, 20 to 30 cm). The area seems to have been thoroughly churned and disturbed: recent material was encountered at deeper levels than sixteenth-century Spanish *majolica* and prehistoric Amerindian pottery (Goggin 1968:40–41).

Trench 2 was dug into the downward sloping town "savannah," close to the junction of Richmond Street and Abercromby Street, where in the 1950s the road cut exposed a midden profile "with considerable shell" (Figure 27; Rouse Papers, St. Joseph field notes). The archaeological remains appeared to be concentrated in a layer of dark brown clayey loam mixed with quartz and schist pebbles, capped by a 10 to 25 cm thick recent fill of red clay and some road material. The occupation layer rested on yellow to orange sterile clay at a depth of 60 to 70 cm. Trench 2 yielded mixed European and Amerindian materials in its upper levels (down to 50 cm depth) and undisturbed prehistoric refuse between depths of 50 and 70 to 80 cm (Goggin Papers, St. Joseph field notes). The European finds can be dated to between the late sixteenth and nineteenth centuries. They include Spanish *majolica* and Middle Style olive jar fragments, delftware, earthenware, chinaware, pipes, pieces of glass as in Trench 1, slate-pencils, a copper ring, iron nails, a knife, and a bullet (Appendices B and C). Amerindian remains, including ceramics of both the Bontour and Mayo complexes, make up most of the material from Trench 2. Apart from pottery (1,364 potsherds), shells—predominantly West Indian Crown conchs (*Melongena melongena*), Tiger lucinas (*Codakia orbicularis*), and River conchs (*Pomacea urceus*)—and animal bones, the prehistoric refuse includes pieces of

Figure 27. Rouse and Goggin's Trench 2 at the St. Joseph 2 site in the process of excavation in 1953. Photograph by Irving Rouse.

quartz crystal, pottery and stone artifacts, flint chips, and a bone point. In Section A2 a human burial was encountered at a depth of 70 to 80 cm, at the very bottom of the deposit, extending out of the left side of the trench (Goggin 1968:40–41, Papers, St. Joseph field notes; Rouse Papers, St. Joseph field notes). The archaeological materials excavated by Rouse and Goggin are at the Yale Peabody Museum (YPM ANT 255532–255926), except for the animal bone fragments, which are curated in the Florida State Museum. The latter collection was analyzed by Wing (1962:4, 45–46, 49; see also Newson 1976:49, 52–54). The shell remains were discussed by Boomert with Peter L. Percharde (Boomert 1985). Finally, in 1981 and 1982 Boomert made an extensive surface collection of Amerindian as well as European pottery, including predominantly Spanish *majolica* and Middle Style olive jar fragments. This collection is at the University of the West Indies (Boomert 1984b:36). Boomert returned to the site in August 2004, noting at that time that most of its area had been asphalted.

Limited amounts of late sixteenth- through eighteenth-century Spanish *majolica* have been collected at St. Joseph (see Appendix B). Rouse and Goggin's excavations yielded in all thirty-two fragments, including pieces of Santo Domingo Blue on White, San Luís Blue on White, Ichtucknee Blue on Blue, Tallahassee Blue on White, Aucilla Polychrome, San Luís Polychrome, Puebla Blue on White, and various undated tin-glazed wares (Goggin 1968:40–41, 201; Deagan 1987:75; Silver and Faber-Morse 2010). The presence of Columbia Plain (1550–1650), noted by Glazier (1982), is unconfirmed. The various *majolica* wares were either directly imported from Spain (Andalusia) or manufactured in Mexico (Lister and Lister 1974). The presence of Bontour ceramics at the St. Joseph 2 site has given rise to the often quoted but erroneous conclusion that pottery of this complex was still manufactured here when the Spanish cleared the grounds for their capital (Rouse 1953:97–98, 110; Cruxent and Rouse 1958–1959, fig. 4; Bullbrook 1961a:18; Goggin 1968:18–19, 40–41; Newson 1976:14, 24, 26, 60, 79; Glazier 1982). In fact, renewed investigation of the stratigraphic situation at the site shows that two Amerindian pottery complexes are present in St. Joseph: Bontour, which appears to be fully prehistoric, and Mayo, which is predominantly or completely historical. The Bontour complex at the St. Joseph 2 site was called the St. Joseph complex by Boomert (1985), a name that should be abandoned. In Trench 2 (the least disturbed deposit of the site), 91.1% of the Amerindian pottery encountered (a total of 4,357 potsherds) was found at a depth of 0 to 30 cm below the present surface. Bontour ceramics were recovered from the bottom of the site (at 70 to 80 cm) up to its surface, whereas Mayo pottery was confined to the 0 to 30 cm stratum, mixed with European finds. Clearly, mechanical admixture is responsible for the blend of Bontour ceramics and Mayo pottery as well as Spanish and English materials in the upper portion of Trench 2. The Mayo earthenware can be assumed to have been the domestic pottery used by the Spanish and *mestizos* of St. Joseph during the sixteenth through eighteenth centuries. They may have obtained these vessels from the loyal Amerindians living in the *encomiendas* and, afterward, the missions of the island. The paucity of *majolica* (less than 1% of the European ceramics from St. Joseph 2) and preponderance of Amerindian pottery in the Spanish capital of Trinidad illustrate the poor living conditions of its inhabitants.

Mayo

The Spanish–Amerindian mission site of Mayo (VIC-17) is in the westernmost portion of the Montserrat Hills, in the southwestern third of Trinidad's Central Range, about 6 km inland from the Gulf of Paria (Figure 28). The site occupies the area around the Roman Catholic church in the center of the village of Mayo. This originated at the location of the former mission in the mid-nineteenth century (Figure 29). To the northwest the mission was connected with the site of another mission village, Santa Ana de Savaneta, the Savaneta 2 site (CAR-4), by the present Mayo Road (which beyond Tortuga is still known as the Indian Trail Road) and passes through a hamlet called Indian Chain. Mayo sits on top of a ridge some 70 m high, about 250 m west of the Mayo River, a right tributary of the Guaracara River. Its surroundings offer a panoramic view to the west across a gently undulating landscape, nowadays predominantly under sugar cane cultivation, as far as the coast of the Gulf of Paria. Originally this part of Trinidad's Central Range was covered with semi-evergreen to evergreen seasonal forest; today its hilly parts toward the east are under cocoa, citrus, and coffee cultivation or, used for household gardens. Geologically, the Mayo area is typified by the calcareous clays, silts, and thin limestone layers of the Brasso Formation, dating from Oligocene to Miocene times. Soils are light, montmorillonitic, nonacidic and freely drained, dominated by the aquentic chromuderts (vertisols) of the Brasso Clay variety. Precipitation amounts locally to 175 to 200 cm per year.

The former mission village, dedicated to Nuestra Señora de Montserrate, was founded by Catalan Capuchins between 1700 and 1705, and existed as a nucleus of the Amerindian population until 1789, when its inhabitants were transferred to the mission of La Anuncíata de Nazaret de Savanna Grande, now the Princes Town 1 site (VIC-6), some 6 km to the south (Philip 1824:166–167; Newson 1976:164–165, 259). In 1730, the Montserrate mission had as many as 354 inhabitants; in 1774, thirty-seven families with 149 "souls" were counted ("Concerning the Island…" [1780]; Carrocera 1979:257). The village plan probably conformed to the general layout of the Spanish–Amerindian missions in colonial America. It would have had houses aligned on the north, south, and west sides of a central square, on the east side of which the church, presbytery, and other communal buildings were situated, such as the *casa de cabildo*, used for village meetings. In 1714, all four mission villages in this part of Trinidad reportedly were provided with such a town hall (Newson 1976:165; Borde 1982:52–53; see e.g., Carrocera 1968, fig. opposite p. 202). In 1769, the church of Montserrate was destroyed by an earthquake, but was soon rebuilt and is described in 1774 and 1785 as a structure of wattle-and-daub covered with palm leaves ("Concerning the Island…" [1780]; Dorta 1967:300). According to testimony from 1785, its single altar had an image of the Virgin Mary wearing a golden necklace with greenstone beads and a crown of silver (Dorta 1967:383–384). In 1774 a visitor to the Montserrate mission called the village "quite neat" ("Concerning the Island…" [1780]). According to Verteuil (1884:299, 430), in

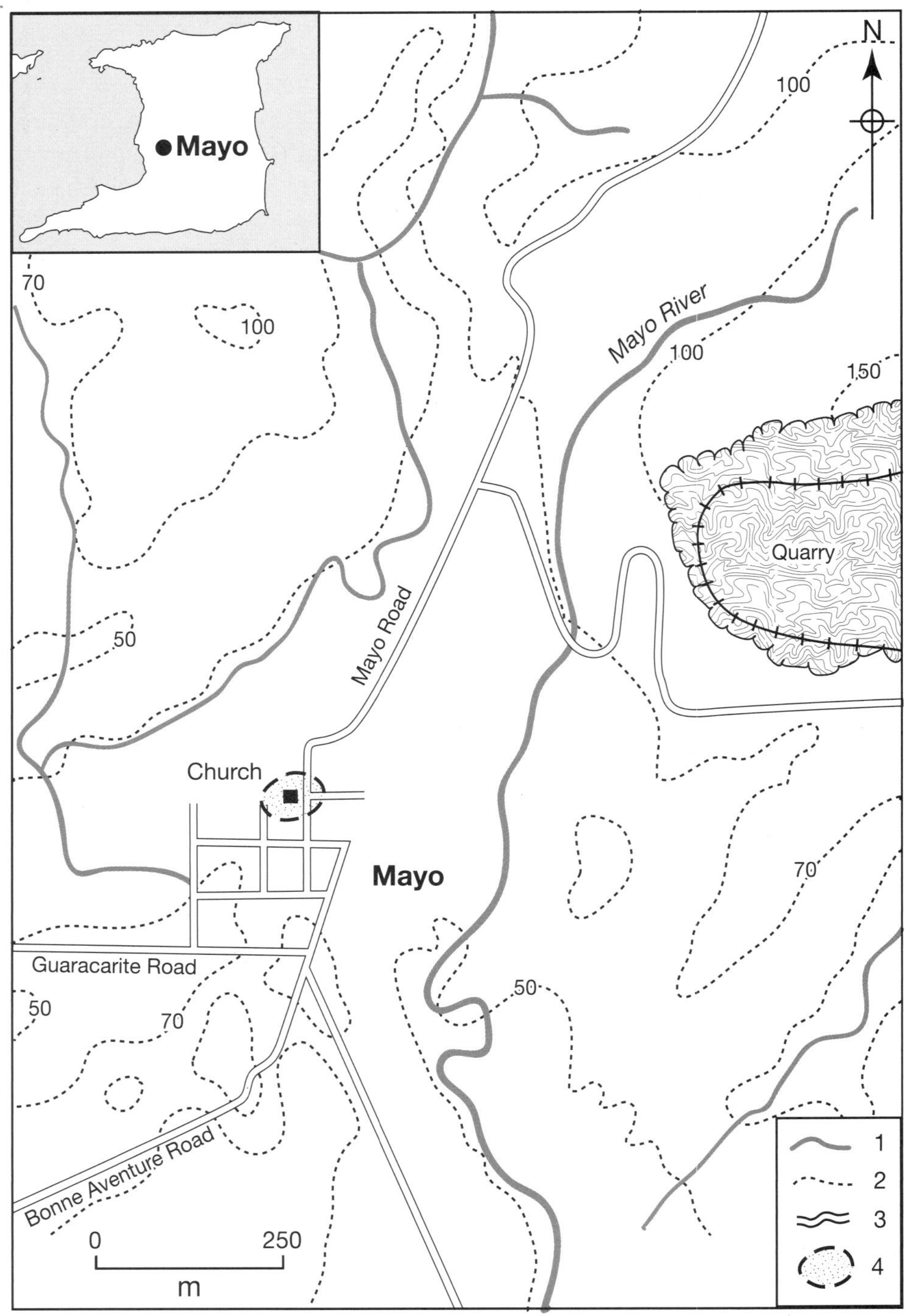

FIGURE 28. The geography of the Mayo (VIC-17) archaeological site in central Trinidad. *Legend:* (1) rivers; (2) 50, 70, 100, and 150 m contours; (3) modern roads; (4) shell midden refuse.

FIGURE 29. The former Mayo Roman Catholic Church and archaeological site in central Trinidad in 1982.

the mid-nineteenth century there still existed "a few vestiges" of the then deserted Montserrate mission. Shortly afterward, in 1867, the village of Mayo was laid out (Kingsley 1889:205–206; Anthony 1988:180–181). It is possible that its (rectangular) plan was based on that of the defunct mission.

The Mayo site was discovered by Rouse and Goggin in August 1953. They encountered shell midden refuse containing Amerindian pottery mixed with European colonial artifacts in a road cut opposite the Mayo church and on the ground next to the church, as well as beneath its raised wooden floor. The following month the Goggins dug three trenches, controlled in 20 cm artificial levels, alongside the church (Figure 30). Trench 1 was divided into three 2 by 2 m units (Sections 1A1 to 1A3), Trench 2 consisted of only one unit (Section 2A1), and Trench 3 was divided into two 2 by 2 m units (Sections 3A1 and 3A2). Trenches 1, 2, and 3 were excavated to approximately 35–50, 30, and 40 cm depths, respectively, until the dark refuse deposit gave way to sterile, hard, yellowish, calcareous clay (Goggin Papers, Mayo field notes; Rouse 1953:100, 110).

The Amerindian earthenware encountered (1,727 potsherds) belongs to the Mayo complex of the Mayoid series. It was found to be mixed with European colonial items in the upper 20 to 30 cm of the deposit (Figures 81 and 84). Apart from pottery, the Goggins recovered many food remains, such as abundant amounts of mollusks, predominantly Tiger lucinas (*Codakia orbicularis*) and Caribbean oysters (*Crassostrea rhizophorae*), and a few animal bones, along with stone and pottery artifacts. The European finds include gunflints, pipestems, English pipe bowls, case

FIGURE 30. Goggin's trenches, alongside the Roman Catholic Church at the Mayo site, in the process of excavation in 1953. Photograph by Irving Rouse.

bottle glass, pieces of lead, a red stone bead, and ironware, including hand-forged "rose head" nails, a musket ball, knives, horseshoes, and door hinges, along with some white-glazed earthenware, chinaware, delftware, and *majolica*. According to Goggin (1968:40), the colonial pottery and artifacts predominantly date from the late-eighteenth and nineteenth centuries and, consequently, post-date the deposition of Amerindian refuse during the existence of the Montserrate mission, that is, before 1789. Only a few pieces of eighteenth-century delftware and some Spanish Middle Style olive jar fragments obviously date from mission times (Goggin Papers, Mayo field notes; see Appendix B). The animal bone materials excavated by Goggin were studied by Wing, who visited the site in July 1959 and made a small surface collection of Amerindian potsherds and food remains (Wing 1962:4–5, 44–46, 49–52; also Newson 1976:52–54, 168). Goggin's finds are curated in the Yale Peabody Museum (YPM ANT 255927–256026), those of Wing are in the collection of the Florida State Museum.

Karen Anderson-Córdova expressed renewed interest in the Mayo site in February–March 1978. She dug a 1 by 1 m test pit controlled in 20 cm artificial levels in the northwest corner of the churchyard and collected from an exposed area behind the church, encountering shell midden refuse down to a depth of 40 cm mixed with animal bones as well as Amerindian ceramics (615 potsherds), stone and pottery artifacts, and European glass, lead, iron, and colonial wares (TTHS Annual Report 1977–1978; Anderson-Córdova 1978; TTHS Newsletter 1978(2)). The last were concentrated in the top level of the test pit and include specimens of eighteenth-century Dutch delftware together with San Luís or Aucilla Polychrome Spanish *majolica*, which can be dated to between about 1650 and 1745 (Kathleen A. Deagan, pers. comm. to Boomert 1982). In addition, Boomert collected Amerindian and European earthenware as well as shell and bone materials from beneath and alongside the Mayo church in December 1981. Both Anderson-Córdova's and Boomert's finds are in the University of the West Indies collection; the shell remains were analyzed by Victoria Jones in November 1984. The site was visited by Saunders and Chauharjasingh in July 1992 and April 1995. They observed shell debris, animal bones, and Amerindian potsherds, as well as European glass, beneath the old church. In 1996 this church, an elegant wooden structure in French provincial style dating from 1898, was dismantled and replaced by a new box-like concrete edifice. Before this construction, in June and July 1996, David D. A. Maharaj and Williams dug a 1 by 1 m excavation, controlled in 10 cm artificial levels, at the site. They recovered a shell midden deposit containing potsherds, charcoal, bone fragments, chert chips, and a few stone artifacts, including a celt fragment and a hammerstone, down to sterile soil at 40 cm below the modern surface, in the upper level mixed with recent European materials (Maharaj 1996). The site was again visited by Saunders and Chauharjasingh in February 1997. At the time they observed that the concrete foundations of the new church had completely covered the midden, although a few potsherds lay scattered around its margins (Saunders and Chauharjasingh 2003:11). The fact that in Mayo, Amerindian refuse dating from the eighteenth century was encountered at the site of the present Roman Catholic church suggests that the latter still stands at its approximate location in mission times. This situation resembles that of several other mission village sites in Trinidad, which in some way or another developed into modern towns (Boomert 1985), notably La Divina Pastora (the Siparia-Pastora Street site [SPA-2]), La Anunciata de Nazaret de Savanna Grande (the Princes Town 1 site [VIC-6]), and Santa Rosa de Arima (the Arima site [SGE-13]).

CHAPTER THREE

The Trinidad Collections

The pre-Columbian cultural chronology of Trinidad as it was devised by Irving Rouse after his 1946 and 1953 excavations on the island has stood the test of time and still forms the backbone of Trinidad's prehistoric cultural sequence. Subsequent work by Harris (1976, 1978, 1991a) and Boomert (1985, 2000) has elaborated and modified Rouse's framework without significantly altering it. Accordingly, several successive Lithic, Archaic, Ceramic, and Historic cultural units can be distinguished in Trinidad's pre-Columbian past, most of which have been named after the sites excavated by Rouse. The various prehistoric complexes that could be defined as a result of his research are discussed here in detail.

Archaic Age Complexes

The individual find of a bifacially chipped, stemmed spearhead at Biche in east–central Trinidad is the only indication that nomadic bands of Lithic hunters and foragers (whose material culture can be classified as belonging to the Canaiman subseries of the Joboid series) already frequented Trinidad as early as about 8000 cal BC, before the island's separation from the mainland (Boomert 2000:47–52). However, the earliest major occupation took place after Trinidad became an island. Archaic (Meso-Indian) hunters, fishers, and food collectors first occupied Trinidad by about 6000 cal BC. Somewhat less than a millennium later the Gulf of Paria and the Columbus Channel gradually attained their present level and configuration. Seafaring and navigating using large dugouts was an integral part of the Meso-Indian cultural heritage. The first movement of Amerindians into the Lesser Antilles was by Archaic Indians from Trinidad and the east Venezuelan littoral by about 5000 to 4000 cal BC. Trinidad's Meso-Indian communities kept up contacts with the Indians of the Paria Peninsula, the west coast of the Gulf of Paria, Tobago, and the Windward Islands until the end of Archaic times (Harris 1976). Today's Warao Indians of the Orinoco Delta and northwest Guyana, who formerly subsisted exclusively on hunting, fishing, and food collecting, may represent the direct descendants of the Ortoiroid Indians of Archaic times. This is suggested also by Warao speech, which is an isolated language unrelated to those of the present horticultural Indians in the region.

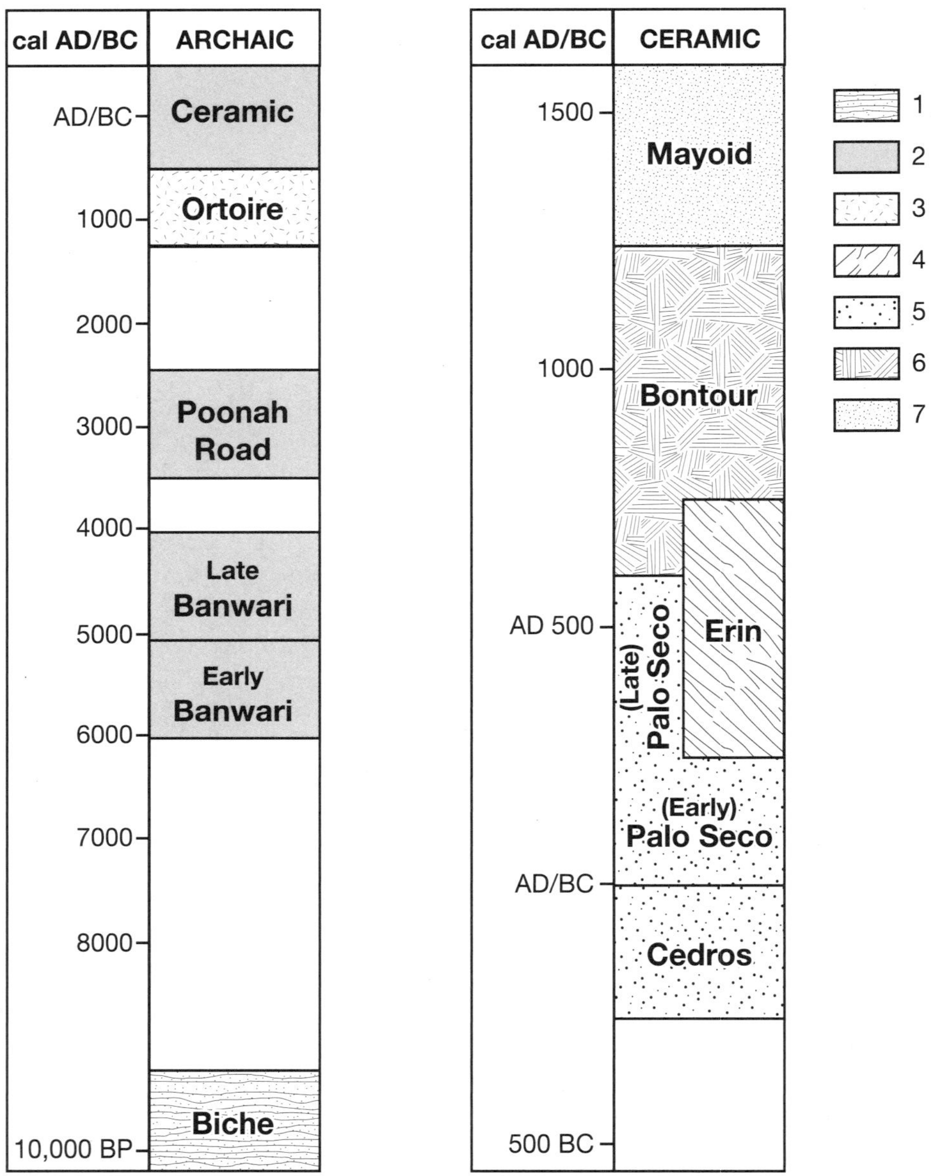

FIGURE 31. Chronology of Trinidad prehistoric through protohistoric cultural complexes according to calibrated radiocarbon dates. *Legend:* (1) Canaiman Joboid; (2) Banwarian Ortoiroid; (3) Ortoiran Ortoiroid; (4) Erinan Barrancoid; (5) Cedrosan Saladoid; (6) Guayabitan Arauquinoid; (7) Mayoid. BP, Before Present (AD 1950). Adapted from Boomert (2000, figs. 7 and 15).

The available series of radiocarbon measurements suggests that the Archaic occupation of Trinidad extended from about 6000 until 950 cal BC (Figure 31). Two subsequent episodes of Archaic settlement in the island can be distinguished: the Early Archaic and Late Archaic periods. The onset of the Late Archaic era has been dated at about 4000 cal BC. In all, twenty-four Archaic sites are known, com-

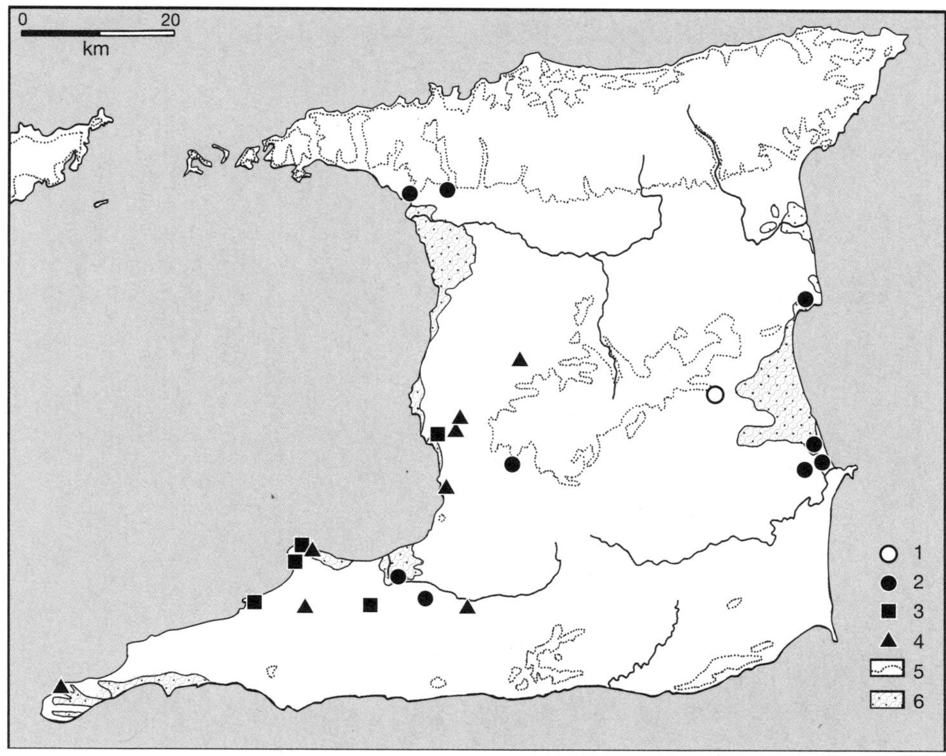

Figure 32. The locations of Lithic and Archaic sites on Trinidad. *Legend:* (1) individual find (Lithic); (2) midden sites (Archaic); (3) flint deposits (Archaic); (4) individual finds (Archaic); (5) 100 m contour; (6) swamps and marshes. Adapted from Boomert (2000, fig. 5).

prising midden sites, flint deposits, and individual finds (Figure 32). They are assigned to a single cultural tradition, the Ortoiroid series, named for the Ortoire site in southeastern Trinidad (Rouse and Allaire 1978). Two Ortoiroid subseries can be distinguished, Banwarian (which encompasses two individual cultural complexes, Banwari Trace and Poonah Road) and Ortoiran. Ortoiran includes only one well-defined complex, Ortoire. In addition, several insufficiently known Archaic sites have been identified that cannot be ascribed to a specific cultural complex within the Ortoiroid series. All Early Archaic sites belong to the Banwari Trace complex, while the Poonah Road and Ortoire complexes can be dated to the Late Archaic period (Boomert 2000:53–91). Rouse's 1953 research involved the excavation of two Archaic sites: St. John, an Early Archaic site that can be assigned to the Banwari Trace complex of the Banwarian subseries, and Ortoire, a Late Archaic site that is the type site of the Ortoire complex of the Ortoiran subseries.

Banwari Trace (Banwarian, Ortoiroid)
Two major shell midden sites, Banwari Trace and St. John, are characteristic of the Banwari Trace complex. Both sites are on the southern edge of the Oropuche Lagoon in southwest Trinidad. Banwari Trace (SPA-28), the type site of the complex,

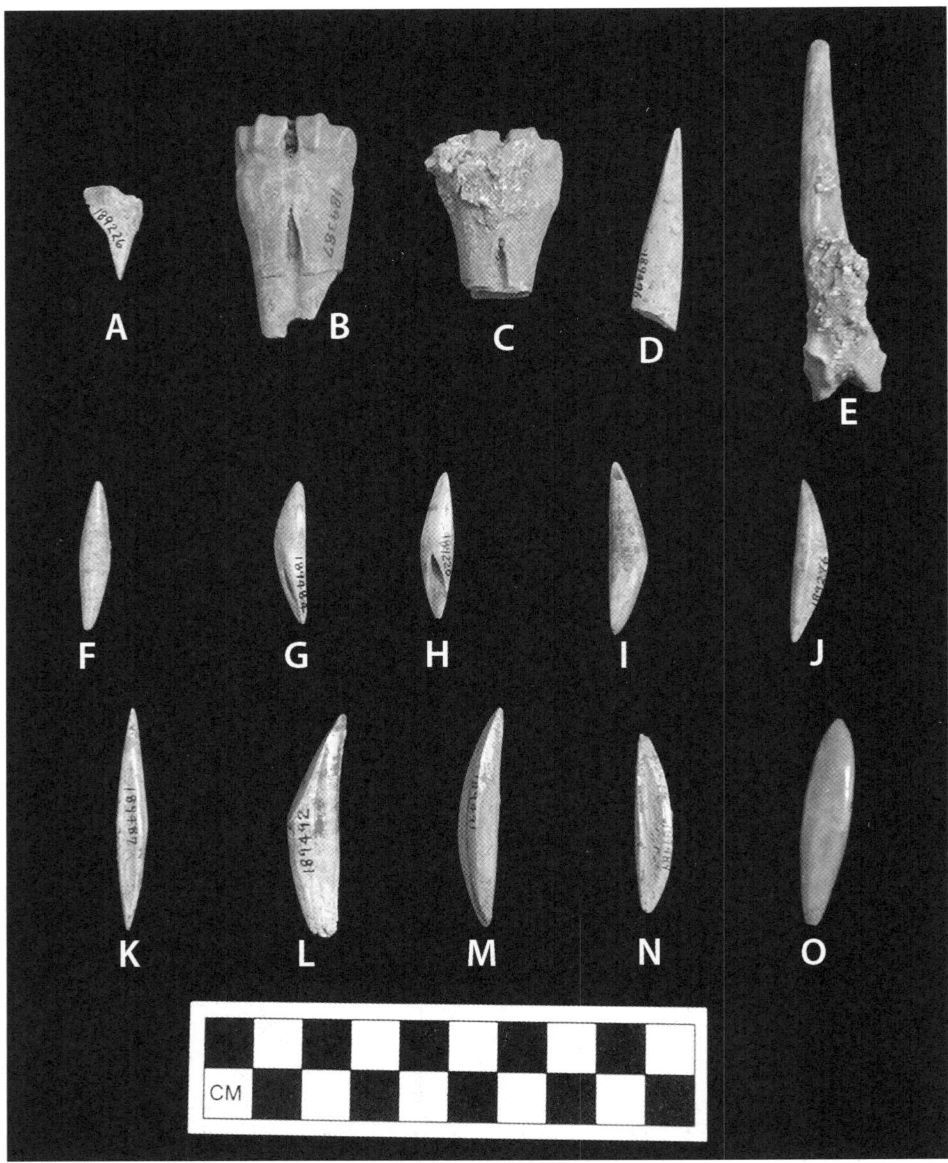

FIGURE 33. Bone artifacts of the Banwari Trace complex from the St. John site (Excavation A). **A,** shark's tooth. **B, C,** cut deer bone. **D,** projectile point. **E,** antler tip. **F–K, M–O,** bipointed fishhooks. **L,** beveled peccary tooth. *Location finds:* A, C, H, A1: 0–20 cm. B, L, M, A4: 20–40 cm. D, A3: 40–60 cm. E, J, O, A3: 0–20 cm. F, A2: 40–60 cm. G, A2: 0–20 cm. I, A4: 0–20 cm. K, A1: 20–40 cm. N, A2: 20–40 cm. *YPM Catalog Nos.:* (A) ANT 189226; (B) ANT 189387; (C) ANT 189221; (D) ANT 189496; (E) ANT 189280; (F, J) ANT 189276 (G) ANT 189484; (H) ANT 189220; (I) ANT 189485; (K) ANT 189487; (L) ANT 189492; (M) ANT 189491; (N) ANT 189489 (O) ANT 189493

was first investigated by Harris from 1969 to 1971. He continued excavating at this site in 2005. The well-dated vertical stratification of Banwari Trace shows a gradual change in the exploitation of the local food resources by the Archaic Indians, which can be correlated with the habitat changes at the Oropuche Lagoon caused by the full submergence that created the Gulf of Paria by about 5100 cal BC (Harris 1973, 1976). Concurrent with the Early Banwari Trace complex, the adaptive strategies of the local Indians are mainly characterized by the hunting of land mammals, fishing for inshore and estuarine species, and collecting freshwater and estuarine shell species. This changes to gathering predominantly marine and brackish mollusks and the prevalence of fishing over hunting. The archaeozoological finds encountered by Rouse at St. John closely reflect the Banwari Trace sequence (see Chapter 4), suggesting that deposition at both sites took place simultaneously. This is indicated also by the ^{14}C date available for (Early) St. John. The shore of the Gulf of Paria must have been considerably farther away from the site than it is today. The Banwari Trace and St. John shell middens can be seen as the refuse deposits belonging to the dwelling sites of a semi-sedentary population characterized by a highly diversified subsistence economy, which exploited a wide range of environmental niches. Perhaps they were the central base camps of these hunters–fishers–foragers, especially the men, who often may have departed temporarily to occupy small camps, quarry sites, or workshops in order to pursue specialized activities.

The Banwari Trace complex shows a highly distinctive cultural assemblage, typically consisting of implements made of stone and bone (Boomert 2000, figs. 9 and 10). The artifact inventory of St. John is closely comparable to that of the Banwari Trace site. It includes objects associated with hunting and fishing, such as bone projectile points used for tipping arrows and fish spears (Figure 33D), a shark's tooth most likely used for similar purposes (Figure 33A), a beveled peccary tooth used as a fishhook (Figure 33L), and bi-pointed pencil hooks of bone intended to be attached in the middle to a fishing line (Figure 33F–K, M–O; Rouse 1960, fig. 4G–I).

A variety of ground stone tools were manufactured for processing plant foods, including blunt or pointed conical pestles (Figure 34A; Harris 1973, fig. 3c–e, 1976, fig. 9), large grinding stones, round to oval *manos* and small mortars for pulverizing red ochre (Figure 34C, D; Harris 1976, figs. 16, 25), and pitted stones or anvils for cracking palm nuts. A highly conspicuous type of *mano*, the so-called side ("faceted") grinder, shows traces of grinding exclusively around the edges. Grooved stone axe heads provided with shovel blades obviously served for the felling of trees and the manufacturing of dugout canoes. Both the Banwari Trace and St. John middens yielded a large variety of small, irregular chips and cores of quartz, chert, and other rock materials made by percussion flaking using the bi-polar technique. This rather unsophisticated stone reducing technology involves resting the core on an anvil stone and striking it with a hammerstone, several of which are known from St. John (Figure 34B, E). Very few flakes seem to have been reworked intentionally by secondary retouch. Clearly, they were intended for various tasks in their unmodi-

FIGURE 34. Stone artifacts of the Banwari Trace complex from the St. John site (Excavation A). **A,** conical pestle. **B, E,** hammerstones. **C, D,** grinding stones. *Location finds:* A, A3: 40–60 cm. B, A1: 40–60 cm. C, A4: 40–60 cm. D, A4: 0–20 cm. E, Excavations of 1936–1938, surface. *YPM Catalog Nos.:* (A) ANT 189437; (B) ANT 189422; (C) ANT 189449; (D) ANT 189313; (E) ANT 189468.

fied state. Smooth antler tips (Figure 33E) may have served a multitude of purposes as well. The function of the several cut ends of deer bone recovered from St. John is enigmatic (Figure 33B, C). Shell artifacts are extremely rare at St. John; one piece may represent a pendant (Harris 1976, fig. 21).

Large amounts of broken and crumbling, often firecracked, soft sandstones have been found at both Banwari Trace and St. John. Most likely these functioned as heating (cooking) stones. Large fish, game meat, and edible tubers may have been cooked or roasted in hearths containing many heating stones that seldom needed replacing. Medium-sized and small fishes were prepared in a similar way or perhaps stewed by dropping hot stones into a gourd, skin, or (waterproof) basket filled with water. Petrographic analysis of the rock types used by the Banwarian peoples for manufacturing implements indicates that local as well as overseas stone materials were exploited. Hard quartzitic rocks made into a variety of milling, pounding, and mashing tools at the Banwari Trace and St. John sites are locally available in large numbers, for instance, on Trinidad's south coast. Similarly, the soft and crumbling sandstones are local, deriving from the littoral zone of the Gulf of Paria, north of the

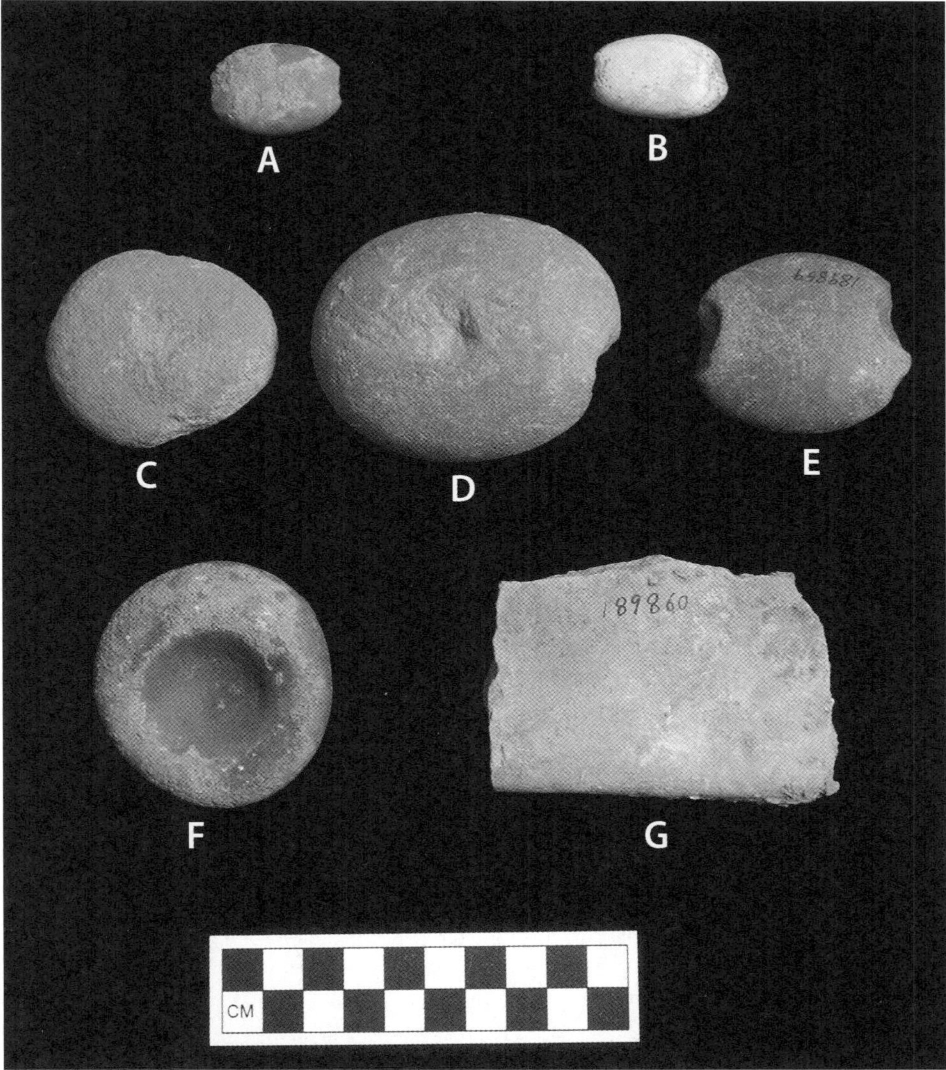

FIGURE 35. Stone artifacts of the Ortoire complex from the Ortoire site (Excavation A). **A, B, E,** net sinkers. **C, D,** anvil-hammerstones. **F,** paint mortar. **G,** grinding stone. *Location finds:* A, A3: 20–40 cm. B, A2: 20–40 cm. C, A6: 120–140 cm. D, A1: 120–140 cm. E, G, A3: 120–140 cm. F, A4: 60–80 cm. *YPM Catalog Nos.:* (A) ANT 189593; (B) ANT 189583; (C) ANT 189871; (D) ANT 189845; (E) ANT 189859; (F) ANT 189711; (G) ANT 189860.

Oropuche Lagoon. The same applies to various low-quality cherts, lumps of red and yellow ochre, pieces of fossilized wood, and pebbles of gray argillite and mudstone. The many limestone fragments encountered at St. John were obviously collected in the Northern Range (Harris 1973, 1976). In contrast, some conical pestles seem to have been made of stone materials from overseas, perhaps from the Windward Islands or Tobago. Two ground stone implements from Harris's 1972 excavations at St. John, made of andesite and leucodiorite, are clearly Tobagonian in origin.

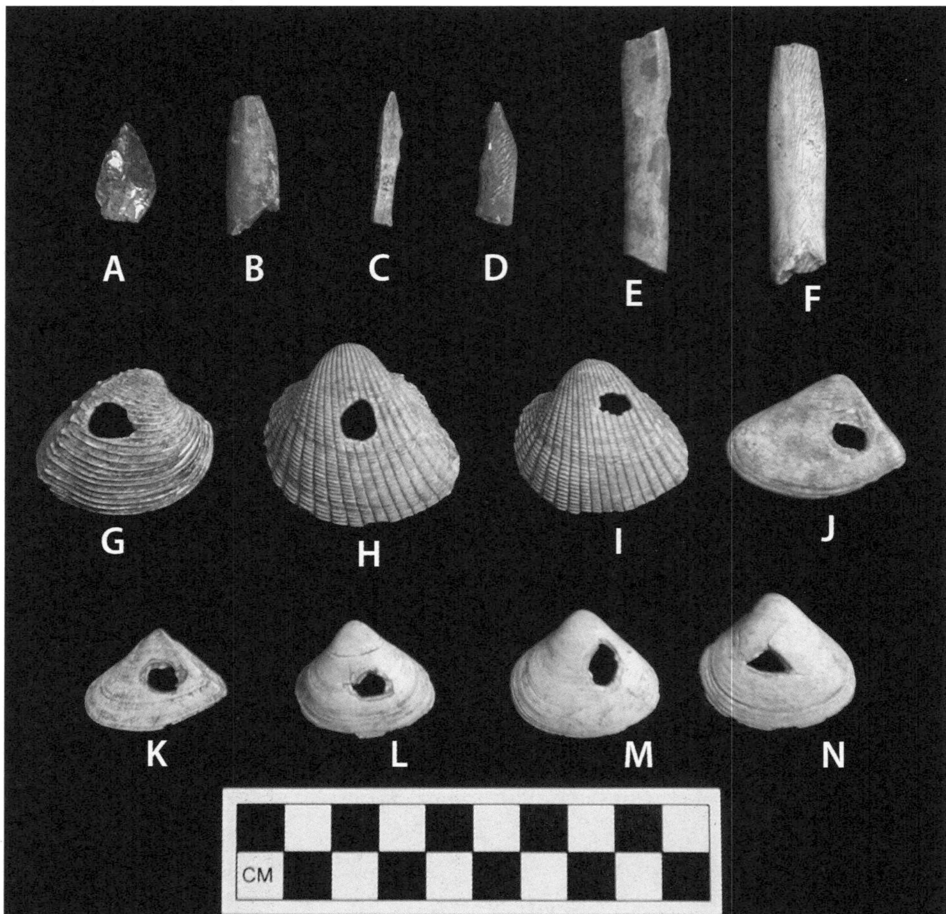

Figure 36. Stone, bone, and shell artifacts of the Ortoire complex from the Ortoire site (Excavation A). **A,** quartz crystal point. **B–F,** bone projectile points. **G–N,** perforated clam shells: (G) *Pitar dione,* (H, I) *Anadara notabilis,* (J, K) *Donax striatus,* (L–N) *Tivela mactroides. Location finds:* A, A4: 100–120 cm. B, unknown. C, A5: 20–40 cm. D, A3: 80–100 cm. E, A5: 100–120 cm. F, unknown. G–I, A6: 20–40 cm. J–N, A2: 20–40 cm. *YPM Catalog Nos.:* (A) ANT 189819; (B) not numbered; (C) ANT 189613; (D) ANT 189763; (E) ANT 189834; (F–N) not numbered.

All this indicates that the Early Archaic inhabitants of Trinidad were skilled canoe builders and competent navigators, showing a definitely maritime inclination.

Ortoire (Ortoiran, Ortoiroid)

The Late Archaic Ortoire complex is characterized by two shell midden deposits, the Cocal 1 (NAR-3) and Ortoire sites, the remnants of which are just north of the estuary of the Ortoire (Guatuaro) River on Trinidad's southeast coast. The food remains recovered from both sites suggest that the Ortoire subsistence strategies closely resembled those of the Banwari Trace complex. The Ortoire and Cocal 1 sites may represent the midden deposits of a central base camp occupied by a semi-sedentary population of hunters–fishers–gatherers who exploited the littoral environment of southeast Trinidad in Late Archaic times. The Ortoire complex is characterized by a cultural assemblage, typically consisting of artifacts of stone and bone (Boomert 2000, fig. 12:6–9, table 4). The Cocal 1 site yielded only a few small pieces of quartz crystal and some hammerstones (de Booy 1917). A more diversified toolkit was excavated by Rouse, at Ortoire. Artifacts related to hunting and fishing include flat, angled projectile points of bone (Figure 36B–F) and small stone netsinkers (Figure 35A, B, E; Rouse 1960, fig. 4K). Most tools were apparently meant for woodworking and the processing of various raw materials and plant foods. Such implements include hammerstones, small stone mortars occasionally showing red coloring due to the grinding of hematite (Figure 35F; Rouse 1960, fig. 4A), pitted anvils for cracking palm nuts (Figure 35C, D; Rouse 1960, fig. 4D), *manos*, and grinding slabs (Figure 35G). Tiny, percussion-flaked chert and quartz crystal chips and cores are most numerous (Rouse 1960, fig. 4B, E, F). Few of them show secondary retouch; one of these specimens may be a projectile point (Figure 36A). Such quartz crystal artifacts may represent exchange valuables since, as still today, pieces of quartz crystal are prized shamanic charms among the Warao Indians of the Orinoco Delta (Boomert 2000:85).

Most ground stone artifacts found at the Ortoire site are made of quartzite found locally in southern Trinidad. Pieces of fine-grained, black dolerite among the rock materials suggest contacts with Tobago. The quartz crystal pieces were probably obtained from Trinidad's Northern Range. The Ortoire site also yielded large quantities of red ochre pebbles and flat, tabular heating stones of local sandstone (Boomert 2000, table 4). The latter specimens are comparable to the cooking stones recovered from the sites of the Banwari Trace complex. Finally, a series of perforated clam shells has been recovered from Ortoire, including *Pitar dione, Anadara notabilis, Donax striatus*, and *Tivela mactroides* (Figure 36G–N). They may have been pendants, fishing lures or net or line sinkers, since it is unlikely that the perforations were made to extract the animal. Clams like these and other bivalve shells were most likely processed either by heating in water using cooking stones or by steaming and baking over hot stones in open hearths. (The inhabitants of Trinidad's south and east coasts still today make a broth of *chipchip* shells by boiling them in water.) In all, three additional Ortoiran shell midden sites are known from elsewhere in southeast Trinidad (Harris 1976; Boomert 2000:86). There is no evidence that the Ortoiran subseries on Trinidad survived as late as the Ceramic age. The contrary seems to have been the case, with the final assemblage of the Manicuaran subseries on Cubagua Island off Venezuela. Here a Late Archaic assemblage has been encountered in association with small amounts of Early Ceramic "trade" pottery, suggesting close contacts between the local hunters, fishers, and food collectors and the intrusive horticulturalists of the Saladoid series.

Ceramic Age Complexes

Trinidad was settled by small groups of pottery-making horticulturalists from the South American mainland by about 300 cal BC. These probably Arawakan-

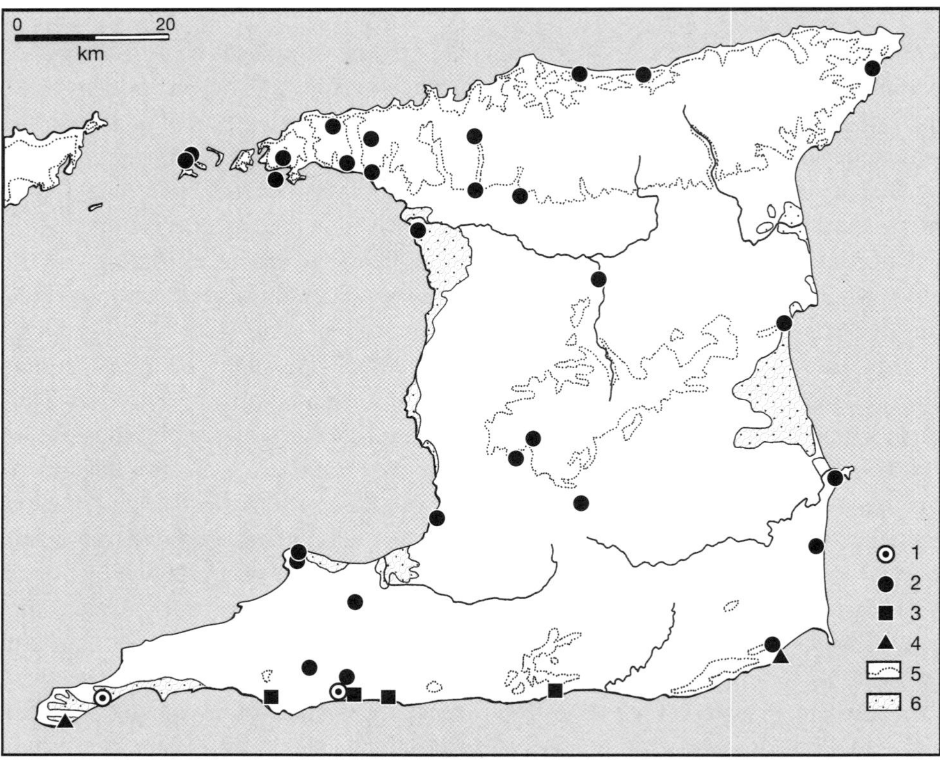

Figure 37. The locations of Early Ceramic sites on Trinidad. *Legend:* (1) Cedros and Palo Seco complexes, Cedrosan Saladoid; (2) Palo Seco complex, Cedrosan Saladoid; (3) Palo Seco complex, Cedrosan Saladoid, and Erin complex, Erinan Barrancoid; (4) Erin complex, Erinan Barrancoid; (5) 100 m contour; (6) swamps and marshes. Adapted from Boomert (2000, fig 16).

speaking Indians, who added the cultivation of bitter cassava and other root crops to the indigenous subsistence economy, probably reached the island from the eastern littoral of Venezuela or the Orinoco Delta. Their cultural tradition is known as the Saladoid series, named after the site of Saladero on the Lower Orinoco. The Saladoid settlement in Trinidad marked the onset of the island's Ceramic age and is characterized by two successive pottery complexes, Cedros and Palo Seco, both considered to belong to the Cedrosan subseries of the Saladoid series (Figure 37). Simultaneously with their occupation of Trinidad, the Saladoid peoples rapidly colonized the West Indian archipelago and by the first centuries AD had established themselves from the South American mainland to as far north as Puerto Rico.

When Rouse developed the first framework of Trinidad's prehistoric cultural chronology, he distinguished six "periods" of Ceramic settlement in the island (Rouse 1947, fig. 1). The Cedros and Palo Seco complexes typify the first four periods in this system. Rouse's fifth period of Ceramic Trinidad is characterized by the Erin complex, belonging to the Erinan subseries of the Barrancoid series. The Amerindians of the Barrancoid series dislodged the Saladoid peoples from the Lower Orinoco by about 800 cal BC. Their cultural influence, which has been attributed to the establishment of a localized interaction sphere, radiated from the mainland to the Saladoid communities of coastal Venezuela, Trinidad, and beyond from the time of Christ onward (Rouse 1983; Boomert 2000, 2009). The (partially overlapping) succession of the Cedros, Palo Seco, and Erin complexes characterizes the Early Ceramic age of Trinidad, Periods I, II, and III of Harris (1978). The onset of the Late Ceramic age, about cal AD 650–800, is marked by the establishment of still another mainland pottery assemblage on the island, the Bontour complex, a representative of the Guayabitan subseries of the Arauquinoid series. It would last until shortly before the end of Trinidad's prehistoric epoch (Figure 31).

The pottery vessel assemblages and earthenware artifacts of Trinidad's various Ceramic age complexes as represented in the Yale collections are discussed separately from the nonpottery artifacts that are described collectively. The descriptive methodology used in the analysis of the pottery vessels follows the format of Rouse and Faber-Morse (1999) in their discussion of the ceramic complexes encountered at the Indian Creek site on Antigua and that used by Boomert (2000) in his analysis of the Early Ceramic (Cedrosan and Erinan) assemblages of Trinidad. The present terminology largely follows that developed by Hofman (1993), Rice (1987), Rye (1981), and Shepard (1968) for the grades of pottery surface treatment, the different vessel shape classes, and the vessel size categories. Accordingly, four types of surface treatment are distinguished: polished, burnished, smoothed, and scraped surfaces. Vessels with height-to-diameter ratios of less than 0.30 are "dishes," those with ratios between 0.30 and 0.50 are "bowls," and those with ratios larger than 0.50 are "jars." Finally, three orifice size categories are recognized. "Small" vessels include those with orifice diameters up to 20.0 cm, "medium-sized" ones are those with orifice diameters from 20.1 to 40.0 cm, and "large" ones are those with orifice diameters exceeding 40.0 cm. The fabric analysis of selected pottery samples of Trinidad's ceramic complexes is discussed in Appendix A.

Cedros (Cedrosan, Saladoid)

Pottery of the Cedros complex is known from two sites on the southern shore of the Cedros Peninsula in southwest Trinidad, Cedros, and Palo Seco (Figure 37). Although the Cedros site yielded predominantly ceramics of the Cedros complex, pottery representative of the transition from Cedros to Palo Seco was encountered in the upper portion of Rouse's Excavation 1 (levels 1 and 2, to a depth of 40 cm below the present surface). At Palo Seco ceramics of the Cedros complex were recovered from the bottom, levels 5 to 7 (80 to 140 cm below the present surface), of Rouse's Excavation 2, while pottery transitional between the Cedros and Palo Seco complexes typifies levels 3 and 4 (40 to 80 cm) of this same trench. Similarly, Section D4 in 1969 yielded Cedros complex pottery in its lowest level, 100 to 125 cm below the present surface (Olsen 1973, 1974:247–249, 253; Faber-Morse 2010).

MATERIAL

Cedros vessels were made by coiling and fired in an open fire. Sherds are typically

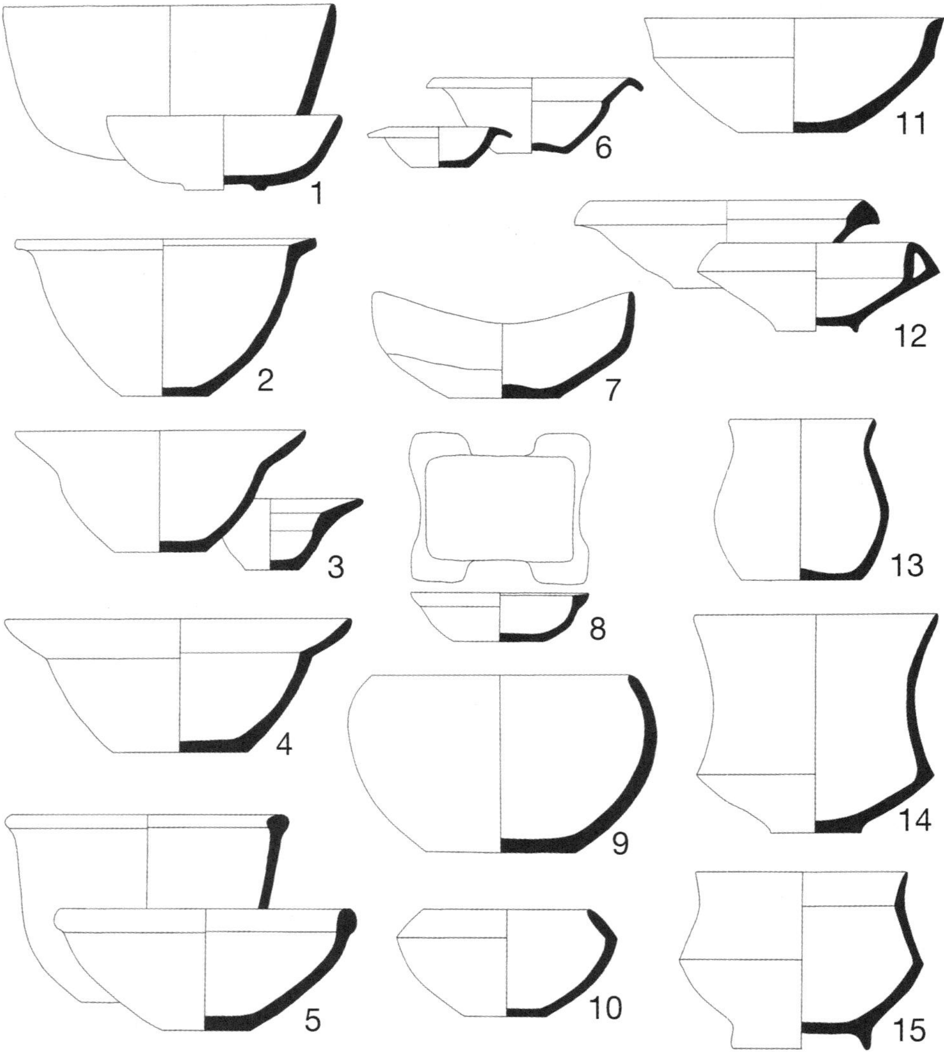

FIGURE 38. Reconstructed Cedrosan Saladoid vessel forms from Trinidad and Tobago (not to scale). Adapted from Boomert (2000, figs. 17–20).

yellowish brown to gray. Surfaces are even, showing careful treatment; most are smoothed or burnished. Some pieces are polished, showing yellowish to orange outer surfaces, possibly due to the application of a specific clay wash before polishing and firing (Kyberg 1976). Complete oxidization prevails. Most vessel walls are relatively thin, 3 to 6 mm, but somewhat thicker cross sections, about 7 to 8 mm, occur as well. Hardness is about 2 to 3 on the Mohs scale. Temper is well mixed with the clay, including predominantly very fine to fine crushed shell and finely ground potsherd (grog), or a combination of these two constituents. Many potsherds con-

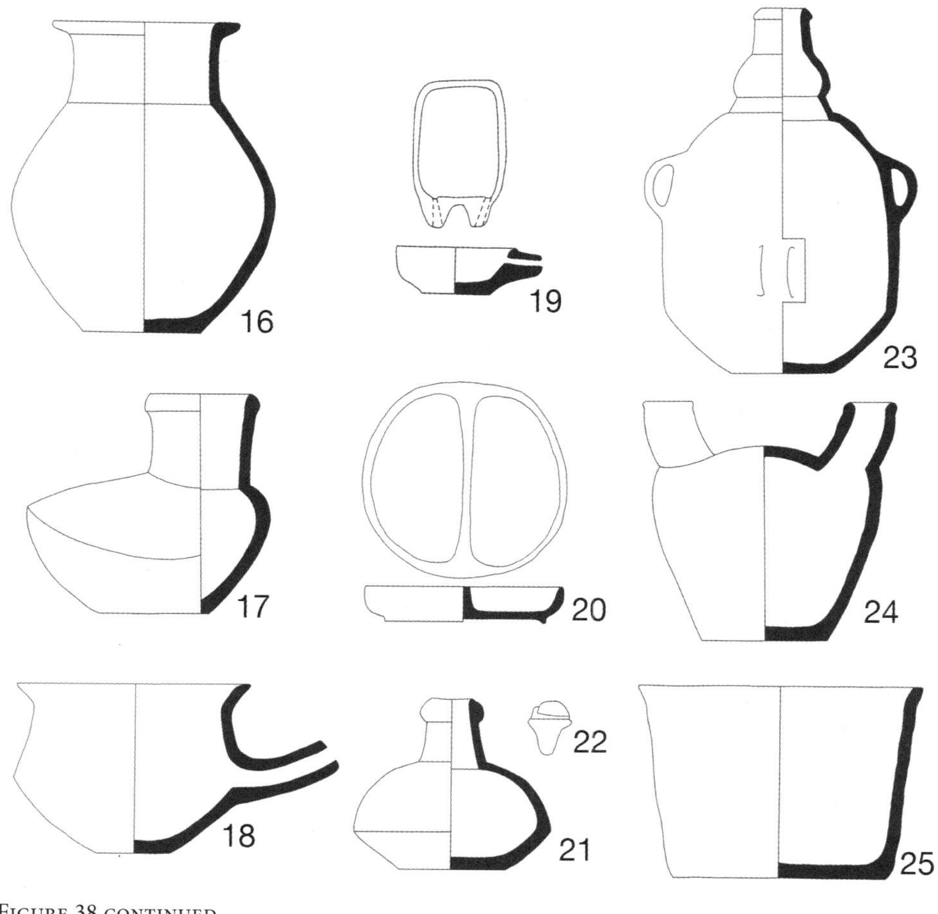

FIGURE 38 CONTINUED.

tain very fine to fine quartz sand particles, which may represent natural additions of the potter's clay; some include a few tiny pieces of what seems to be burned tree bark, possibly *caraipé* (Petersen 2004). Griddles (baking plates) are typically tempered with abundant amounts of crushed shell.

SHAPE

Vessel Shapes. According to the 287 identifiable rim sherds and twenty-one base fragments found at the Cedros site in 1946 (see Boomert 2000, table 7), the common vessel shapes of the Cedros complex encompass Trinidad and Tobago Saladoid Vessel Forms 1, 2, 3, 9, 13, 14, and 21/22, while Forms 4, 5, 6, 8, 10, 11, and 15 are rare (Figure 38; Boomert 2000:132–139, figs. 17–22). Vessels are typically small to medium-sized. Unless otherwise indicated, direct rims are typical, including interiorly or exteriorly thickened or beveled, tapering, and T-shaped forms, often showing flattened lips. Forms 1 (28.6%), 3 (27.9%), and 14 (13.2%) are the most frequent

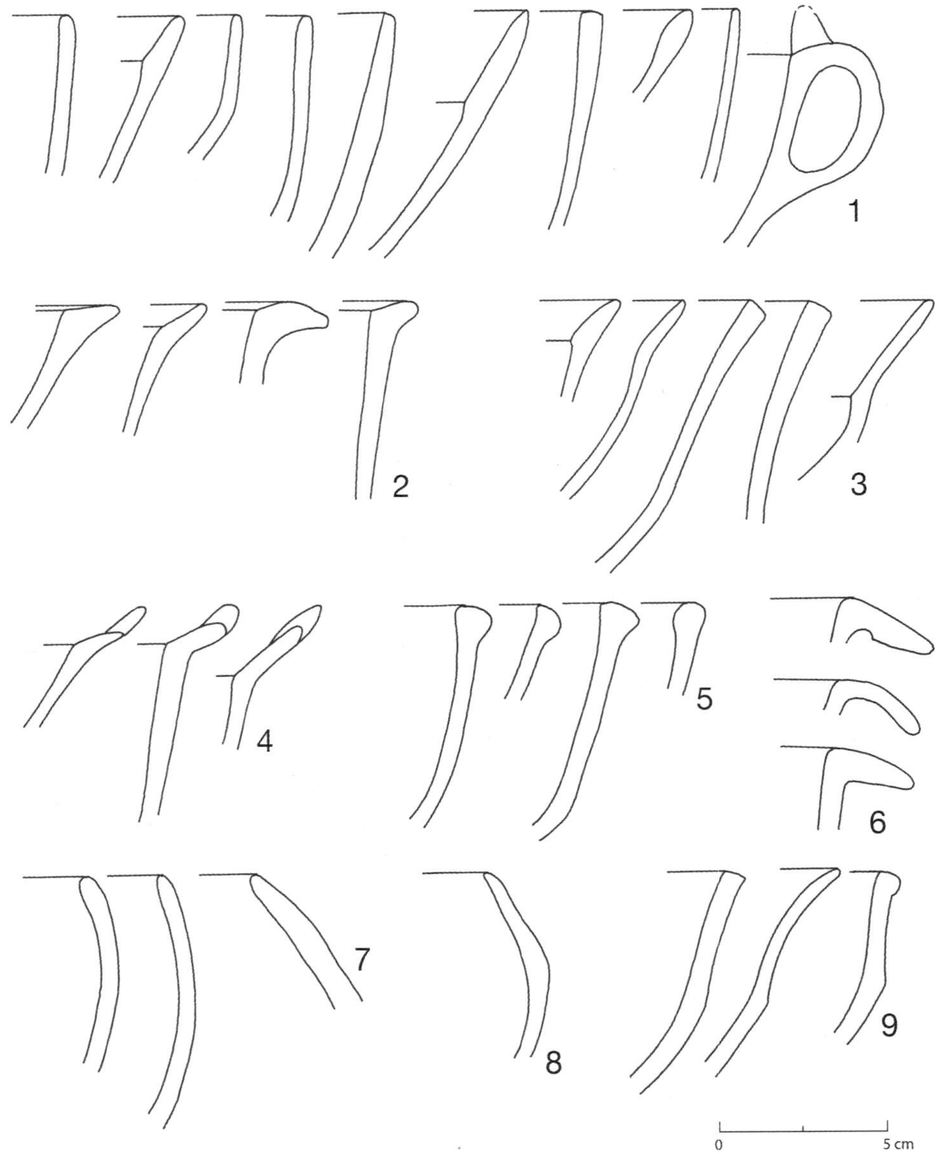

FIGURE 39. Cedros complex vessel forms and rim types from the Cedros and Palo Seco (Excavation 2) sites; (1) TT Form 1; (2) TT Form 2; (3) TT Form 3; (4) TT Form 4; (5) TT Form 5; (6) TT Form 6; (7) TT Form 9; (8) TT Form 10; (9) TT Form 11.

vessel shapes, followed by Forms 2 (9.1%), 9 (5.6%), 21/22 (5.2%), and 13 (3.8%). The common shapes encompass: (1) the simple, unrestricted bowl, dish, or jar of Form 1 (Figure 39:1); (2) the unrestricted bowl with horizontally flanged rim and probably inflected contours of Form 2 (Figure 39:2); (3) the inverted bell-like bowl with independent restricted orifice and inflected or composite contours of Form 3, which may be round or oval in cross section (Figure 39:3); (4) the simple, restricted bowl or jar of Form 9 (Figure 39:7); (5) the jar with independent restricted orifice and inflected contours of Form 13 (Figure 40A:1); (6) the "keeled" bowl or jar with unrestricted orifice and composite contours, showing a concave profile above its corner point, of Form 14 (Figure 40A:2); and (7) the bottle-like jar with independent restricted orifice and exteriorly thickened rim and stopper, destined to cover its opening, of Forms 21/22 (Figure 40A:4, 6).

Two vessel shapes are rare, including the unrestricted bowl with wide, interiorly flattened to concave, flange-like rim and inflected or composite contours of Form 4 (Figure 39:4), and the unrestricted bowl with composite contours and nearly vertical walls above its corner point of Form 11 (Figure 39:9). Four other rare vessel shapes occurring in the upper levels of the Cedros site seem to be transitional to the Palo Seco complex, including: (1) the unrestricted bowl or jar with exteriorly thickened rim and inflected to composite contours of Form 5 (Figure 39:5); (2) the bowl with independent restricted orifice, concavo-convex flanged rim and inflected or composite contours of Form 6 (Figure 39:6); (3) the simple, unrestricted dish with squarish to rectangular cross section of Form 8; and (4) the "biconical" bowl with dependent restricted orifice and composite contours of Form 10 (Figure 39:8). One of these transitional Cedros–Palo Seco vessel shapes, Form 15, has been encountered exclusively in the lower portion of Excavation 2 at Palo Seco. This is a "necked" jar with an independent restricted orifice and composite to complex contours (Figure 40A:3). In addition, the 2005 excavations at the Cedros site yielded the fragment of a simple, unrestricted bowl provided with an Y-shaped rim profile (Figure 40A:5). It was recovered from Pit B1, level 9 (40 to 45 cm below the present surface). This bowl type, thus far undefined for Trinidad, closely resembles Vargas's Vessel Form 27 encountered at Early El Cuartel and El Mayal in the east Venezuelan coastal zone (Vargas 1979:124, 131–132). It has been registered as Trinidad and Tobago Saladoid Vessel Form 26.

The various Cedros vessel classes can be assigned to several functional categories. Vessel shapes probably intended for storage of dry goods or food preparation without heating include Forms 9, 10, and 11, while Forms 13, 14, 15, and 21/22 were most likely used for the storage of liquids or as traveling canteens. Cooking vessels comprise probably Forms 1, 5, and 13, while Forms 1, 2, 3, 4, 5, 6, 8, and 26 may have been used for drying or displaying various kinds of foods and as serving vessels, perhaps primarily for serving cassava beer (Boomert 2000:136–139, 304–307; Faber-Morse 2007).

Base Forms. Four base forms can be identified (Boomert 2000, fig. 22:10): flat (A and D), annular and outflaring or "ring shaped" (B), and concave (C). Bases A, B, and C show unmodified basal angles. Base D is pedestaled; it is rare and occurs only in the transitional phase between Cedros and Palo Seco. Base B is the most frequent form (52.4%).

Handles. Handles encompass predominantly vertical, D-shaped strap handles; they are often peg-topped (Figures 39:1, 40A:3).

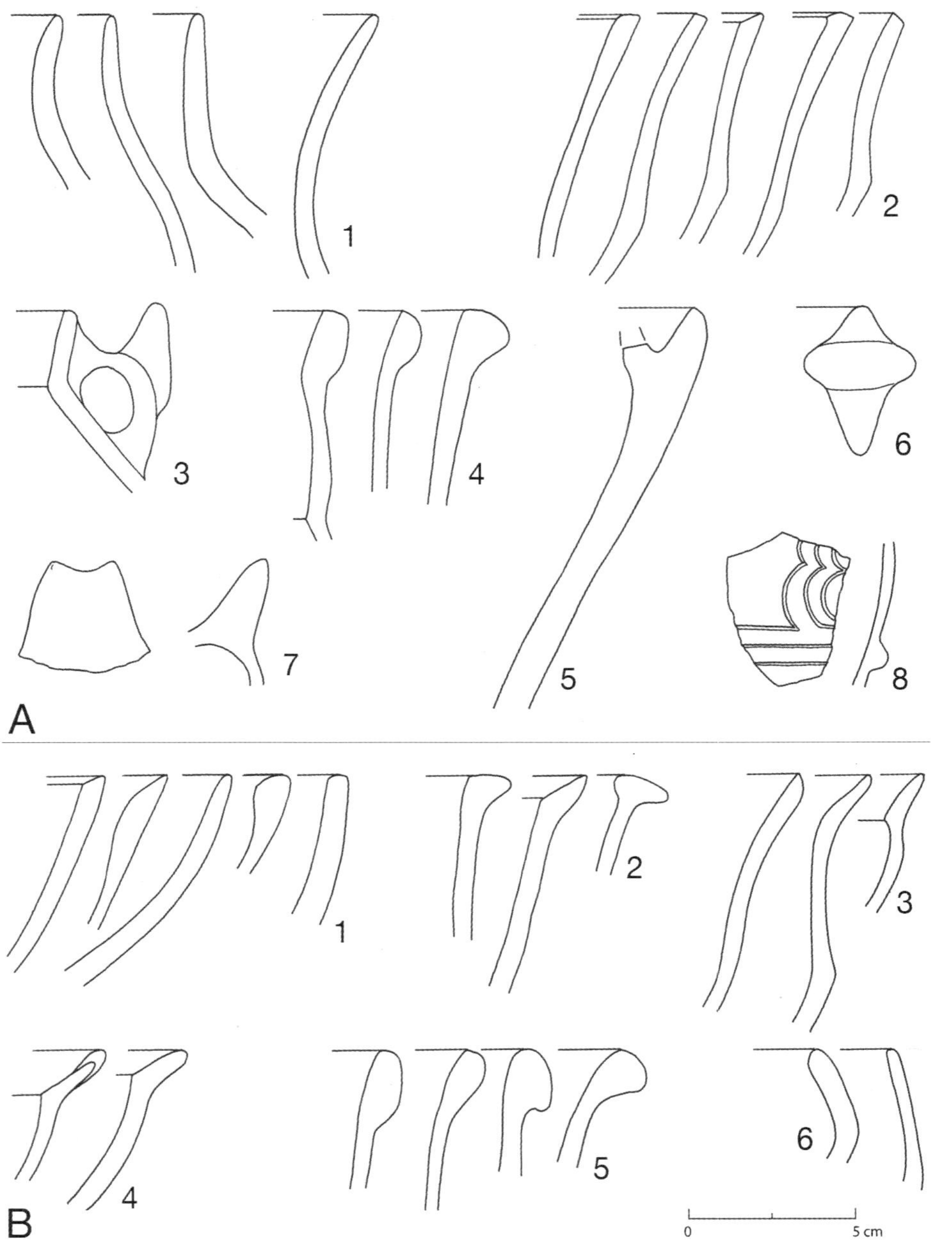

FIGURE 40. Cedros and Palo Seco complex vessel forms and rim types. **A,** Cedros complex, from the Cedros and Palo Seco (Excavation 2) sites. (1) TT Form 13; (2) TT Form 14; (3) TT Form 15; (4) TT Form 21; (5) additional TT Form; (6) TT Form 22. *Decorated pottery:* (7) trapezoid side lug; (8) incised and simple modeled potsherd. **B,** Palo Seco complex, from the Palo Seco (Excavations 1 and 2), Erin (Excavations 1 and 2) and Quinam (Excavations 1 and 3) sites. (1) TT Form 1; (2) TT Form 2; (3) TT Form 3; (4) TT Form 4; (5) TT Form 5; (6) TT Form 9.

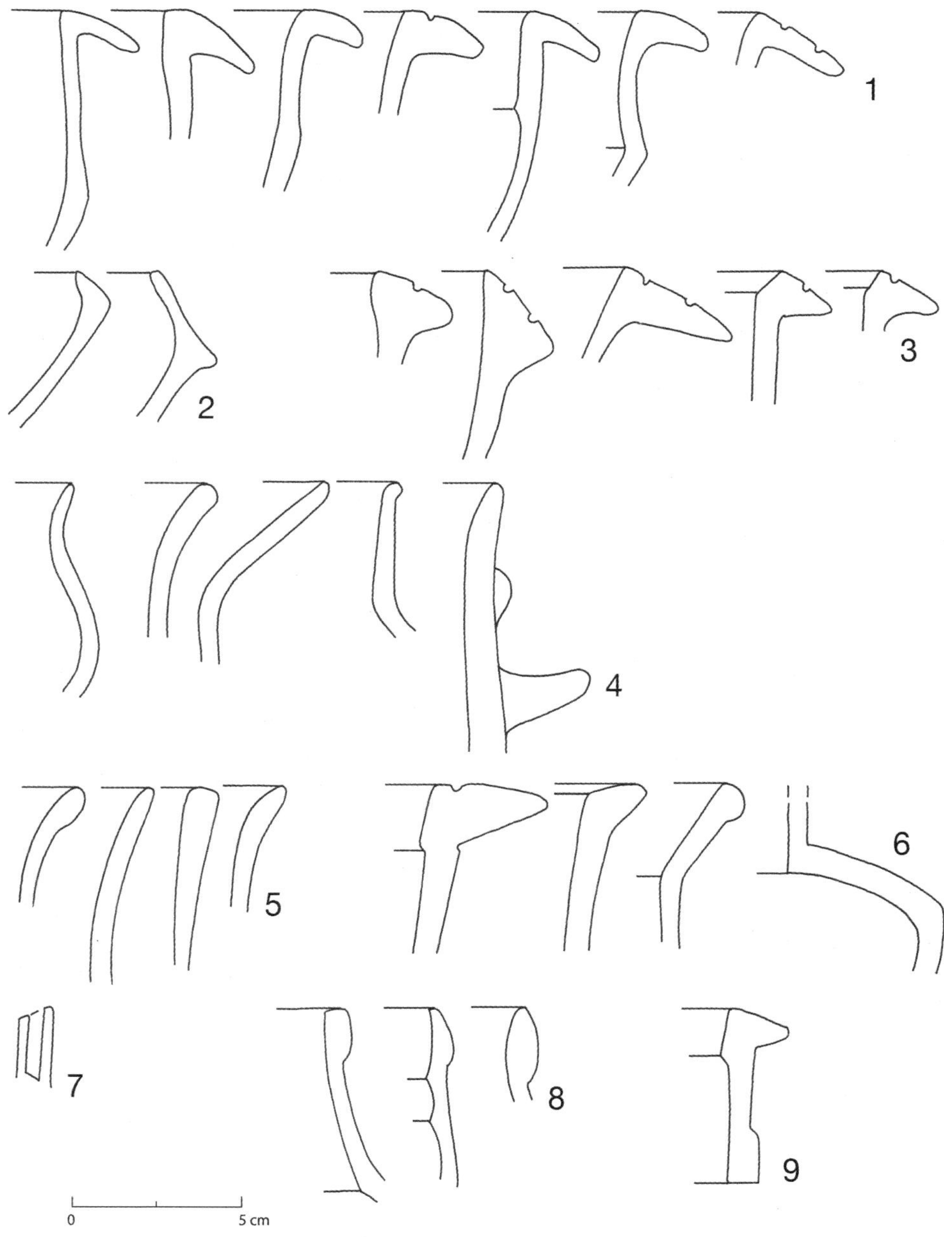

FIGURE 41. Palo Seco complex vessel forms and rim types from the Palo Seco (Excavations 1 and 2), Erin (Excavations 1 and 2) and Quinam (Excavations 1 and 3) sites. (1) TT Form 6; (2) TT Form 10; (3) TT Form 12; (4) TT Form 13; (5) TT Form 14; (6) TT Form 16; (7) TT Form 19; (8) TT Form 21; (9) pot stand.

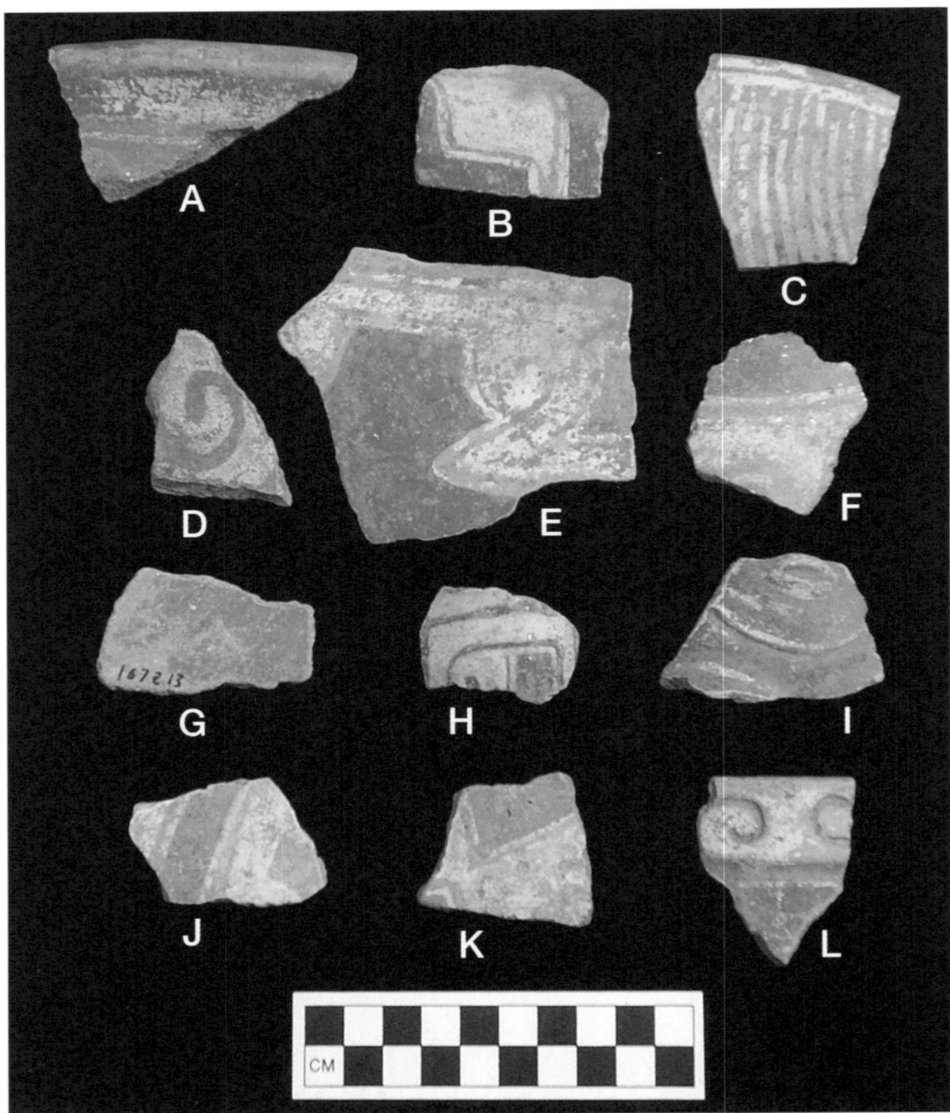

FIGURE 42. Painted pottery of the Cedros complex from the Cedros site (Excavation 1). *Location finds:* A–D, G, K, A3: 20–40 cm. E, A4: 20–40 cm. F, I, J, A6: 40–60 cm. H, A5: 20–40 cm. L, A6: 0–20 cm. *YPM Catalog Nos.:* (A) ANT 167360; (B) ANT 167266; (C) ANT 167269; (D) ANT 167221; (E) ANT 167455; (F) ANT 168365; (G) ANT 167213; (H) ANT 167612; (I) ANT 168415; (J) ANT 168352; (K) ANT 167212; (L) ANT 166968.

DECORATION

In all, 580 potsherds from the 1946 Cedros excavations are decorated (Boomert 2000, table 8). Five decorative techniques can be distinguished: painting, incision, punctation, simple modeling, and complex modeling. Painting is most frequent (51.0%), followed by incision (35.7%), and modeling (22.8%). Punctation is rare (0.3%); it is restricted to the transitional phase between Cedros and Palo Seco.

Painting. The Cedros potters used red-to-orange, white, and (rarely) black pigments. All painted motifs were applied before firing. Deep red pigment was most likely derived from red ochre (hematite), of which several lumps were found at the Cedros site. Kaolin clay was clearly the source of white pigment, and carbon (soot) or some other organic substance that of the black pigment. The inner surfaces of some potsherds show a thick black coating that may have been achieved by burning tree resin under the inverted container, as is still done today by the Amerindians of the Guianas to achieve a watertight surface. Red and white-on-red (WOR) painted motifs are typical, including 56.1% and 29.0% of the painted potsherds, respectively (Figures 42A–F, H, J, K and 49H, K). Scrolls and hourglass designs, achieved by partially scraping off the white slip to reveal the underlying red pigment, occur most frequently. Such red-colored geometric areas are often bounded by thin, white painted and plain lines. Polychrome red, white, and black motifs are rare (Figure 42G, I, L). Characteristic but similarly rare designs include red painted crosshatches (Figure 49H), black-filled incised lines (Figure 42L), and red painted vessel rims. The latter motif is predominantly found in the transitional phase between Cedros and Palo Seco.

Incision. Zone-incised-crosshatched (ZIC) motifs—fine-line incised, zoned crosshatches bounded by deeper and wider lines (see Figures 43A–K, M, 48I, and 49C, L)—are characteristic (45.4%). Spiral and crozier designs, wavy lines as well as multiple, parallel semi-circles and similar motifs consisting of rectilinear lines are commonly incised with a broader stylus (Figures 40A:8, 42H, L, 43F, H, L, 44C–E, G, K, L, 49G, I, J). The ZIC motifs seem to have been applied when the clay was leather hard and, consequently, they were engraved rather than incised (Faber-Morse 2007).

Simple modeling. This decorative technique includes punctated rim pellets, pairs of which are occasionally provided with connecting bars (Figures 42I, 43B, C, F, H–I, M, 44B, C, F, H–J, and 49B, C, E, F, I). Triangular side lugs appear to be applied to D-shaped strap handles (Figure 40A:7).

Complex modeling. A wide variety of modeled-incised, geometric, and anthropozoomorphic head lugs, often surmounting D-shaped strap handles, have been encountered at the Cedros complex deposits of the Cedros and Palo Seco sites (Figures 44A, 45A–J, 48I, and 49A, D). Semi-spherical, solid, or hollow, human-like *adornos* with concave backs and highly conventionalized features consisting of several punctated nubbins are typical. True modeling is rare. Hollow mammiform lugs and zoomorphic *adornos* also occur. The animal species represented are difficult to determine; birds are easiest to recognize and one long-snouted animal somewhat resembles a coatimundi. Another head lug perhaps represents a bat.

OTHER KINDS OF CLAY ARTIFACTS

Apart from pottery vessels, two ceramic artifacts are known from the Cedros

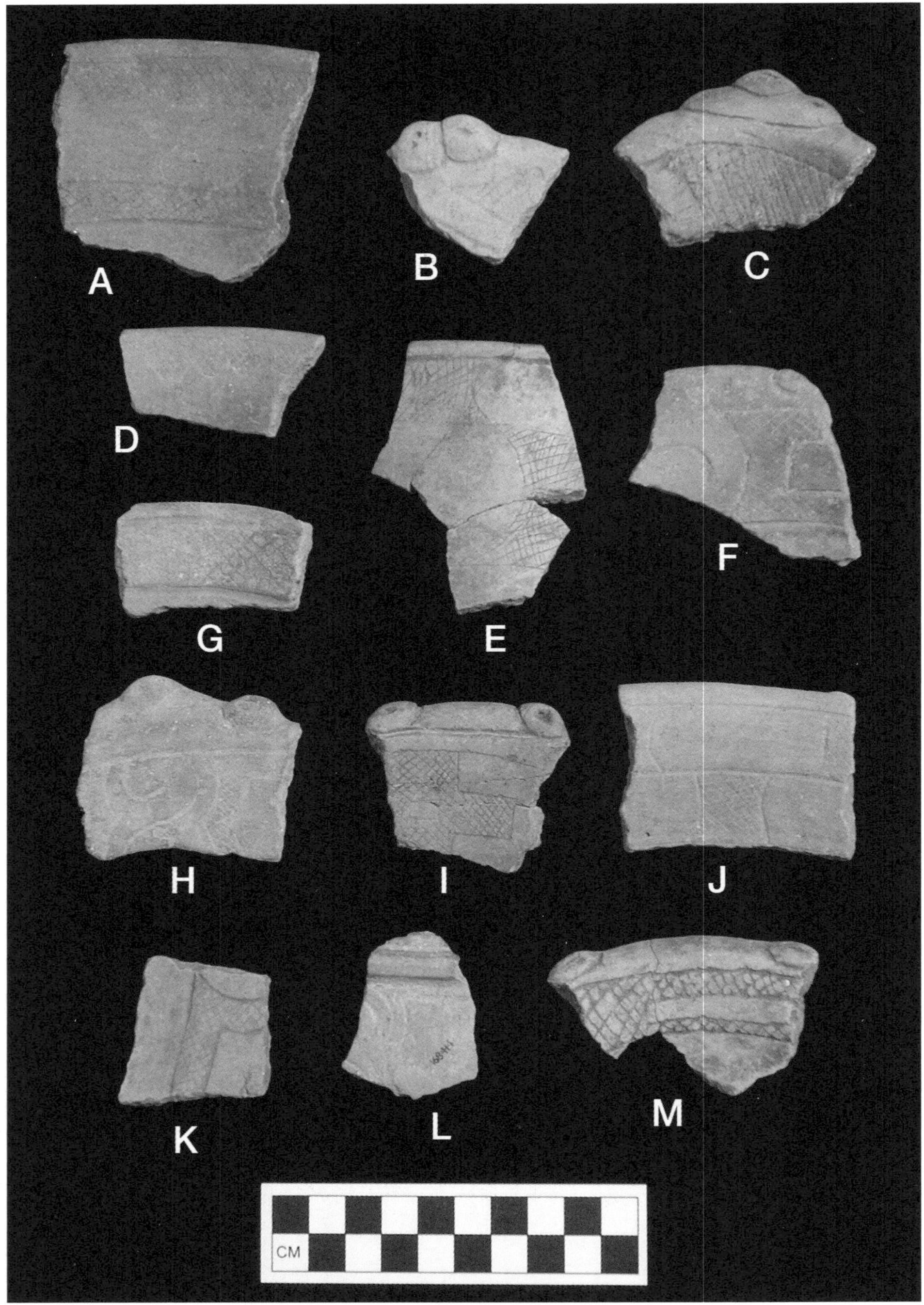

FIGURE 43. Zone-incised-crosshatched pottery of the Cedros complex from the Cedros site (Excavation 1). *Location finds:* A, K, M, A4: 20–40 cm. B, A5: 40–60 cm. C, A1: 0–20 cm. D, L, A6: 40–60 cm. E, A2: 0–20 cm. F, A5: 20–40 cm. G, J, A3: 20–40 cm. H, A2: 20–40 cm. I, A3: 0–20 cm. *YPM Catalog Nos.:* (A) ANT 167516; (B) ANT 168318; (C) ANT 166254; (D) ANT 168498; (E) ANT 166374/166375; (F) ANT 167687; (G) ANT 167341; (H) ANT 167180; (I) ANT 166556; (J) ANT 167343; (K) ANT 167453; (L) ANT 168414; (M) ANT 167517/167518.

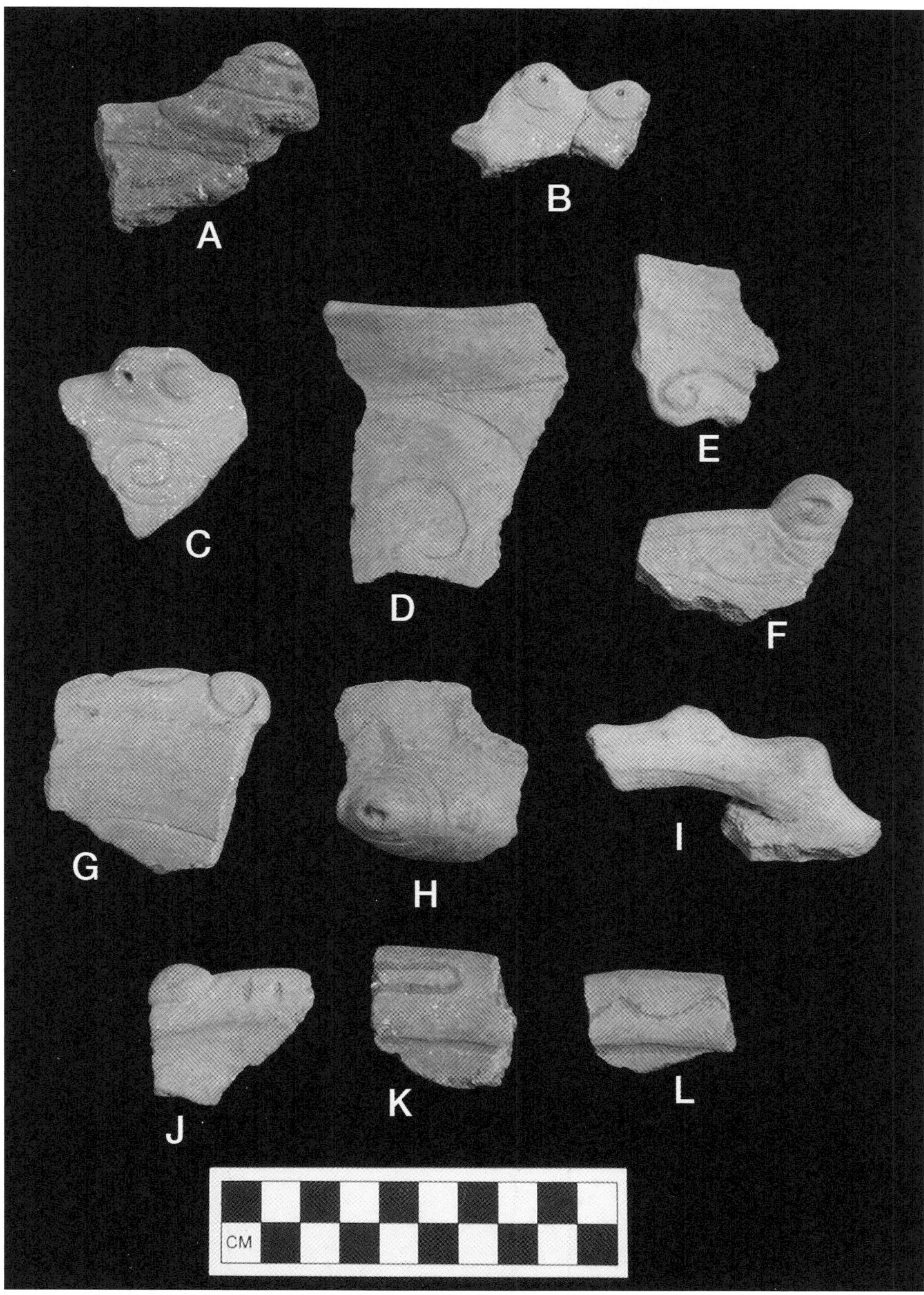

FIGURE 44. Incised and modeled-incised pottery of the Cedros complex from the Cedros site (Excavation 1). *Location finds:* A, K, L, A2: 0–20 cm. B, H, A6: 40–60 cm. C, A3: 20–40 cm. D, A6: 20–40 cm. E, J, A5: 40–60 cm. F, Testpit, 20–40 cm. G, A4: 40–60 cm. I, A6: 0–20 cm. *YPM Catalog Nos.:* (A) ANT 166390; (B) ANT 168504/168505; (C) ANT 167209; (D) ANT 167824; (E) ANT 168032; (F) ANT 168744; (G) ANT 168185; (H) ANT 168416; (I) ANT 166959; (J) ANT 168030; (K) ANT 166373; (L) ANT 166377.

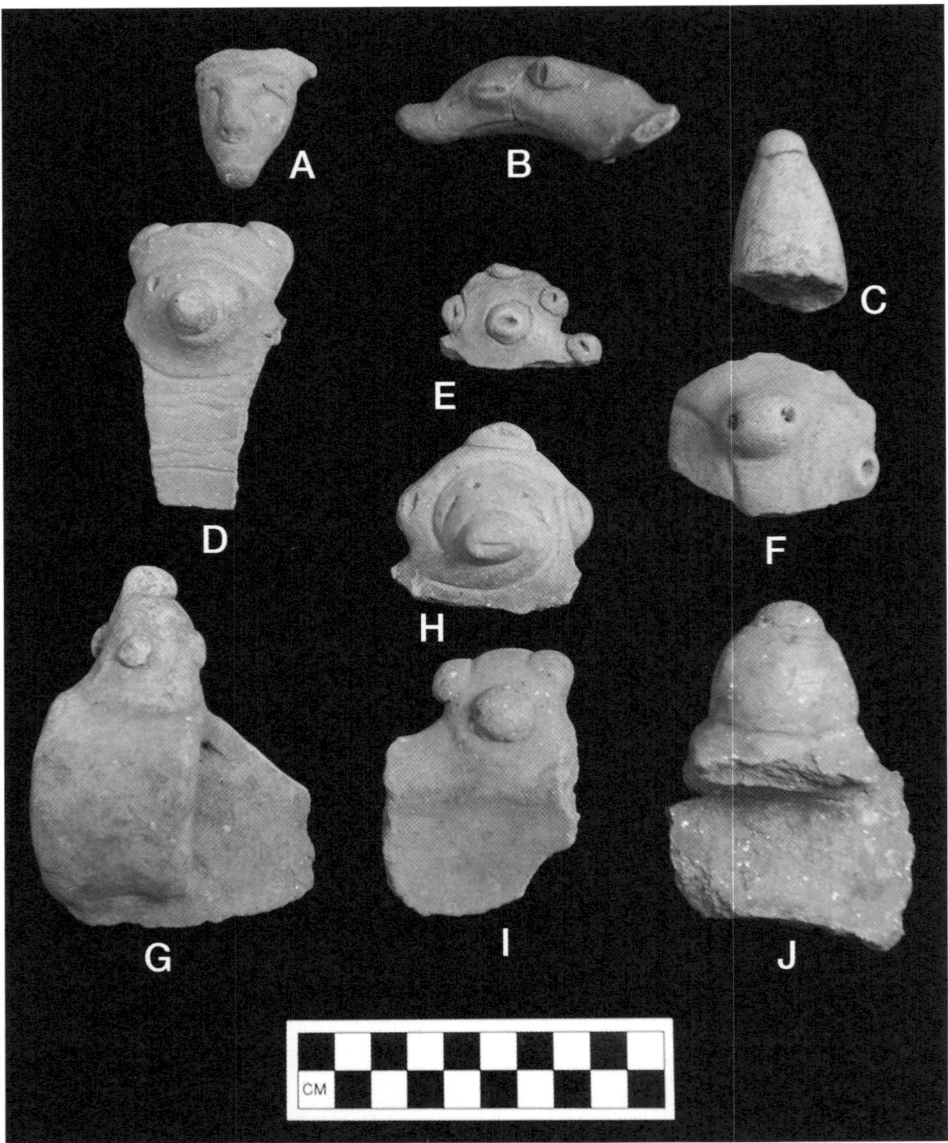

Figure 45. Modeled-incised pottery of the Cedros complex from the Cedros site (Excavation 1). *Location finds:* A, A4: 40–60 cm. B, A4: 0–20 cm. C, A6: 0–20 cm. D, Testpit, 60–80 cm. E, A3: 20–40 cm. F, G, A4: 20–40 cm. H, A6: 60–80 cm. I, A2: 20–40 cm. J, A5: 20–40 cm. *YPM Catalog Nos.:* (A) ANT 168204; (B) ANT 166680; (C) ANT 166971; (D) ANT 168772; (E) ANT 167369; (F) ANT 168624; (G) ANT 167529; (H) ANT 168691; (I) ANT 167181; (J) ANT 167700.

complex: griddles and pottery cylinders. The griddles include thick-walled platters provided with flat, unmodified rims and specimens showing thickened, upturned rims. This indicates that probably both discs of cassava bread and cassava pellets were processed next to other foodstuffs. In addition, pottery cylinders are associated with Cedros pottery, both at the Cedros and Palo Seco sites.

Palo Seco (Cedrosan, Saladoid)

Settlement sites, camp and bivouac sites, and individual finds of the Palo Seco complex have been encountered in all parts of Trinidad. They are especially well distributed in the southwestern part of the island, Mayaro, the Central Range, and the western portion of the Northern Basin and the Northern Range (Figure 37). In all, thirty-four sites are known, including twenty-eight single-component sites. Several Palo Seco midden deposits have yielded human burials. The sites of Chagonary, Erin, Palo Seco, La Lune 1, and Quinam, all on Trinidad's southwest coast, and Manzanilla 1, on the eastern littoral of the island, are multicomponent deposits, typified by Palo Seco pottery and either Cedros or Erin and Bontour materials (Boomert 2000:145–154; Faber-Morse 2010). Palo Seco, the type site of the complex, yielded Palo Seco pottery throughout Rouse's Excavation 1 and in the upper portion of Excavation 2. It is accompanied by small quantities of Erin complex "trade" sherds and a single piece of the Arauquinoid Bontour complex in the upper part, levels 1 to 4 (0 to 80 cm), of Excavation 1. Section D4 excavated at Palo Seco by Rouse, José Cruxent, and Fred Olsen in 1969 yielded ceramics of the Palo Seco complex throughout most of the deposit, levels 1 to 4 (0 to 100 cm). Palo Seco pottery characterizes both trenches dug by Rouse at the Erin site in 1946. In addition, Excavation 1 yielded substantial amounts of Erin materials in its upper and middle portions, levels 1 to 8 (0 to 160 cm), and some Bontour complex "trade" pottery in its top section, levels 1 to 5 (0 to 100 cm), whereas rare amounts of Erin and Bontour ceramics typify the upper part, levels 1 to 2 (0 to 40 cm) of Excavation 2. At Quinam, finally, Rouse recovered ceramics of the Palo Seco complex from his Excavations 1, 2, and 3. It is associated with substantial amounts of Erin pottery throughout the deposits of Excavations 1 and 2, and Erin "trade" sherds in the upper part, levels 1 to 4 (0 to 80 cm), of Excavation 3. In addition, the top and middle portions, levels 1 to 5 (0 to 100 cm), of Excavation 1 yielded a few Bontour "trade" pieces.

MATERIAL

Palo Seco vessels were made by coiling and fired in an open fire. The pottery is quite variable in temper and manufacture. Moderately thick (6 to 12 mm), somewhat coarse and soft ceramics predominate, although relatively thin and fine pottery still occurs. Hardness is 2 to 3 on the Mohs scale. Surfaces are burnished to smoothed and less even than in the Cedros complex. A few Palo Seco ceramics are characterized by a typically very dense and compact light-colored gray texture not showing any clearly discernible temper particles. Extremely well-polished sherds with yellowish to orange outer surfaces due to the application of a clay wash before polishing and firing continue from Cedros times. However, most pieces are yellowish brown to gray or orange. Fine, crushed shell temper and medium to coarse grog temper, or a combination of both, predominate at the sites of Trinidad's south coast. Elsewhere fine to medium-coarse quartz sand or siltstone particles were used as temper. Exclusively grog-tempered vessels are soft and somewhat crumbly, with badly sorted and unevenly distributed temper particles. Temper grains are

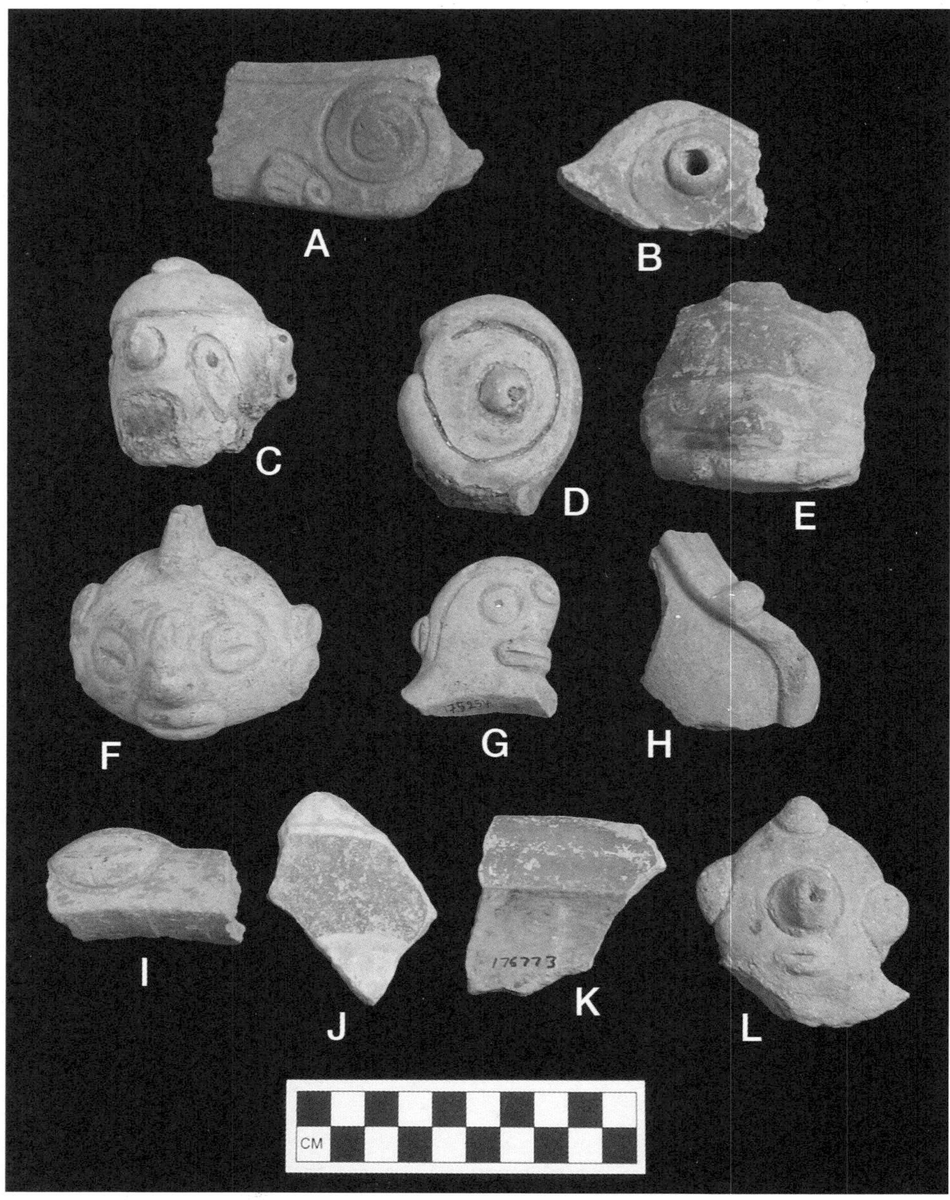

Figure 46. Painted and modeled-incised pottery from the Palo Seco site (Excavation 1). **A, E,** Erin complex. **B–D, F–L,** Palo Seco complex. *Location finds:* A, A4: 0–20 cm. B, A5: 0–20 cm. C, D, A3: 20–40 cm. E, A6: 20–40 cm. F, A2: 40–60 cm. G, A1: 40–60 cm. H, A4: 40–60 cm. I, A3: 80–100 cm. J, A6: 80–100 cm. K, A3: 100–120 cm. L, A3: 120–140 cm. *YPM Catalog Nos.:* (A) ANT 174439; (B) ANT 174538; (C) ANT 174893; (D) ANT 174871; (E) ANT 174052; (F) ANT 175320; (G) ANT 175254; (H) ANT 175482; (I) ANT 176501; (J) ANT 176911; (K) ANT 176773; (L) ANT 176897.

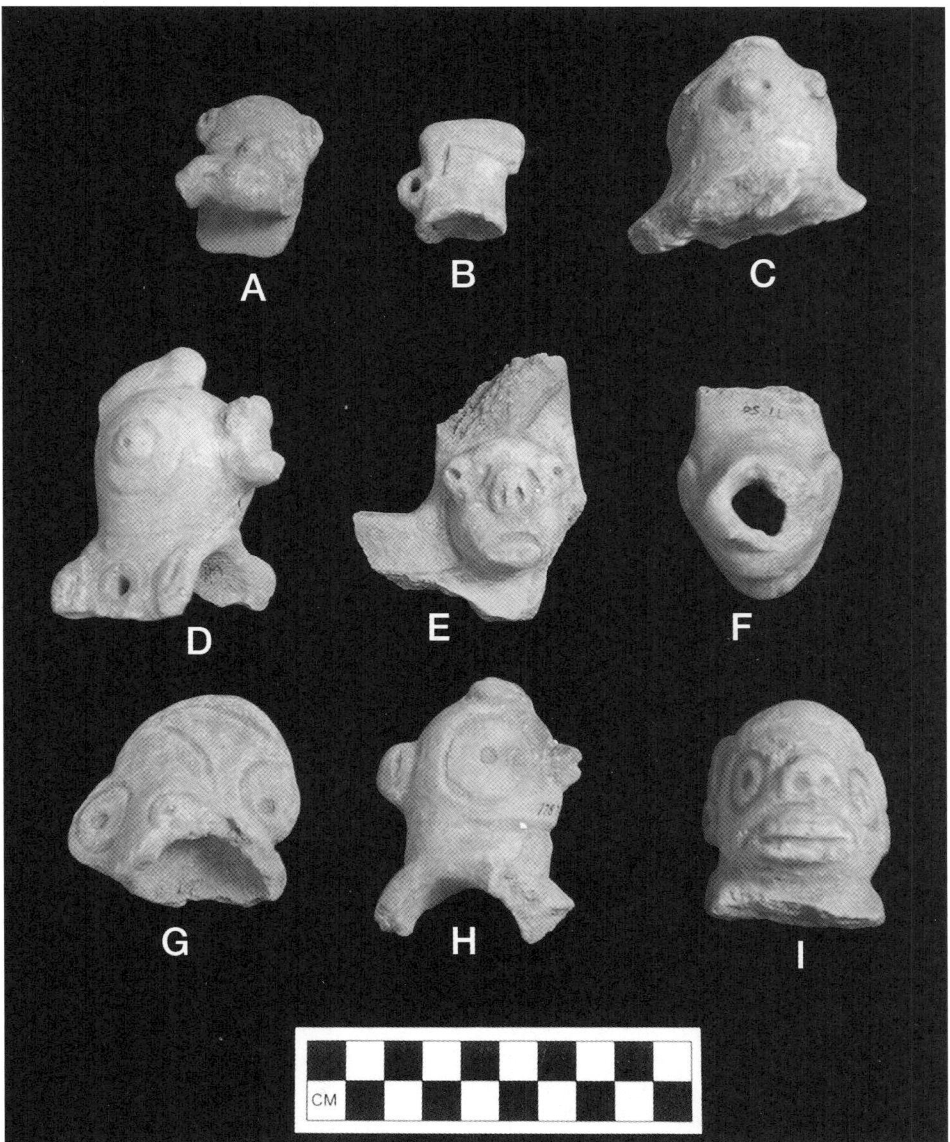

FIGURE 47. Modeled-incised pottery from the Palo Seco site (Excavation 2). **A–E, G–I,** Palo Seco complex. **F,** Erin complex. *Location finds:* A, G1: 0–20 cm. B, F, G3: 0–20 cm. C, E, G2: 0–20 cm. D, I, G2: 20–40 cm. G, G1: 20–40 cm. H, G4: 20–40 cm. *YPM Catalog Nos.:* (A) ANT 177115; (B) ANT 177353; (C) ANT 177254; (D) ANT 178074; (E) ANT 177253; (F) ANT 177354; (G) ANT 177753; (H) ANT 178749; (I) ANT 178076.

sometimes protruding. Some potsherds show heavily corroded surfaces. These pieces most likely are from vessels used for fermenting cassava beer, since the erosive damage can be attributed to the effect of acidic manioc juices. Griddles are tempered with grog or, most frequently, abundant amounts of crushed shell, just as during Cedros times.

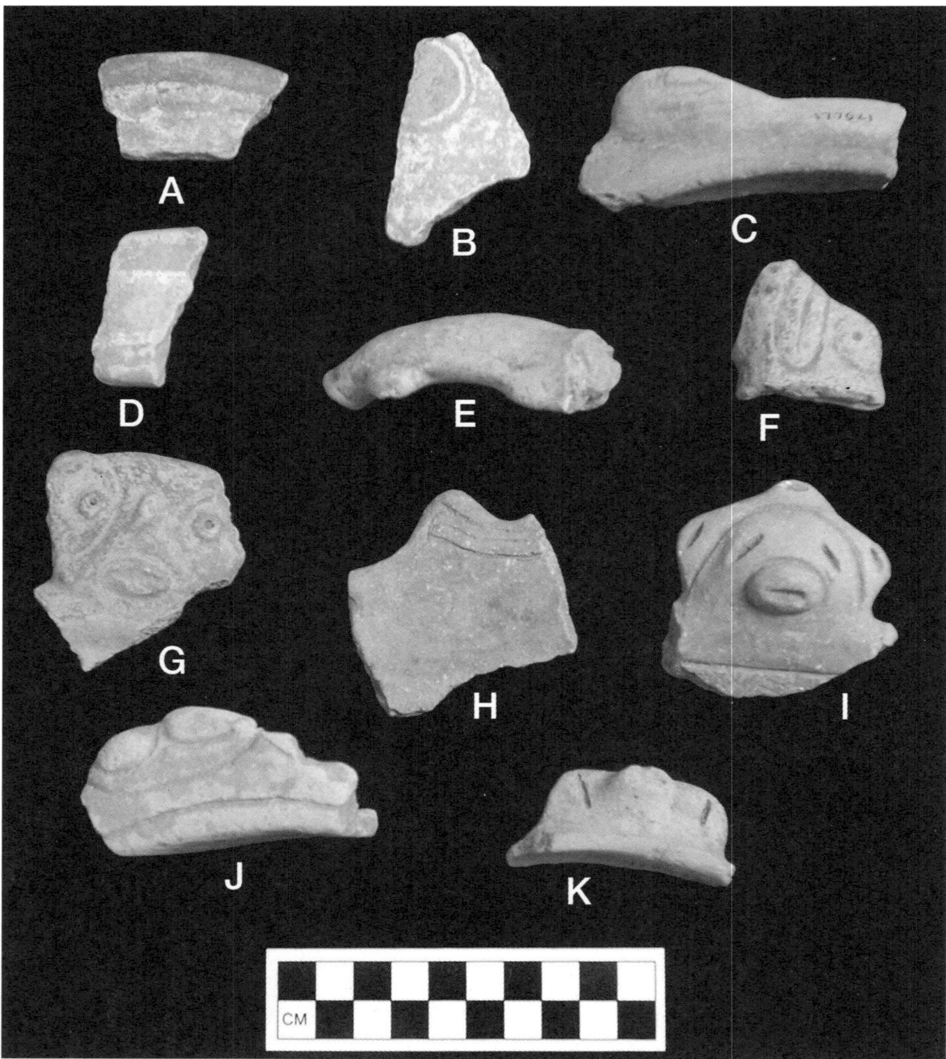

FIGURE 48. Painted and modeled-incised pottery from the Palo Seco site (Excavation 2). **A–H, J, K,** Palo Seco complex. **I,** transition between the Cedros and Palo Seco complexes. *Location finds:* A, D, H, G3: 20–40 cm. B, G, G4: 20–40 cm. C, E, F, G1: 20–40 cm. I, G3: 40–60 cm. J, G2: 20–40 cm. K, G4: 40–60 cm. *YPM Catalog Nos.:* (A) ANT 178237; (B) ANT 178398; (C) ANT 177621; (D) ANT 178386; (E) ANT 177761; (F) ANT 177749; (G) ANT 178746; (H) ANT 178365; (I) ANT 179328; (J) ANT 178009; (K) ANT 178737.

SHAPE

Vessel Shapes. All seven common Cedros vessel shapes and seven rare forms (Forms 1–6, 8–11, 13–15, and 21/22) continue to be used during the Palo Seco complex (Figure 38; Boomert 2000, tables 5, 6, figs. 17–20). However, their frequency of use changes markedly as the flanged bowl of Form 6 (Figure 41:1) and the biconical bowl of Form 10 (Figure 41:2) become common vessel shapes. Indeed, Form 6 represents the most typical bowl form during Palo Seco times whereas, conversely,

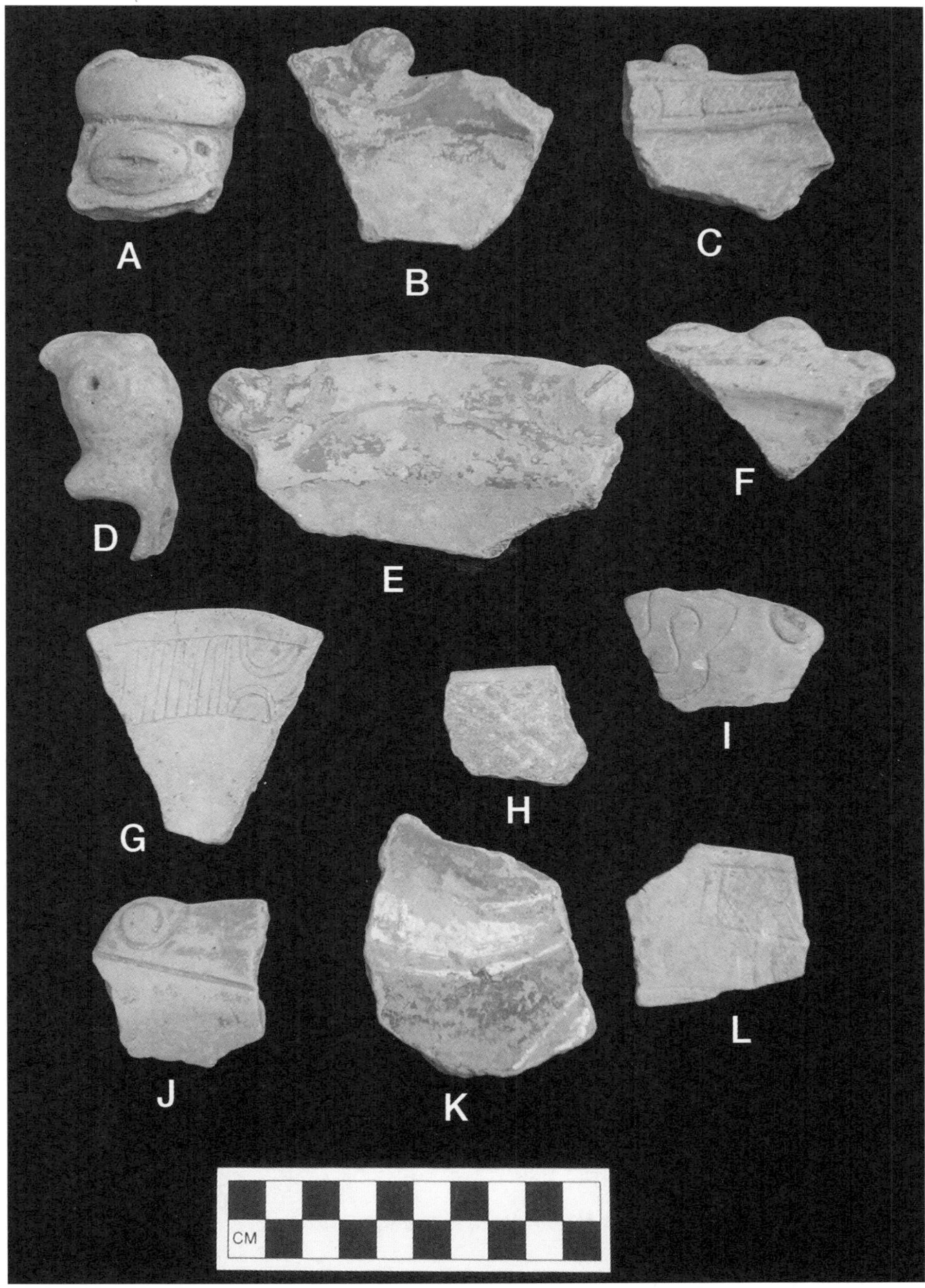

Figure 49. Painted, incised, and modeled-incised pottery from the Palo Seco site (Excavation 2). **A–E,** transition between the Cedros and Palo Seco complexes. **F–L,** Cedros complex. *Location finds:* A, D, E, G1: 40–60 cm. B, G4: 40–60 cm. C, G3: 40–60 cm. F, G, J, G4: 60–80 cm. H, G5: 80–100 cm. I, G5: 120–140 cm. K, L, G5: 60–80 cm. *YPM Catalog Nos.:* (A) ANT 179120; (B) ANT 179449; (C) ANT 179320; (D) ANT 179543; (E) ANT 179448; (F) ANT 179687; (G) ANT 179772; (H) ANT 180042; (I) ANT 180131; (J) ANT 179793; (K) ANT 179814; (L) ANT 180033.

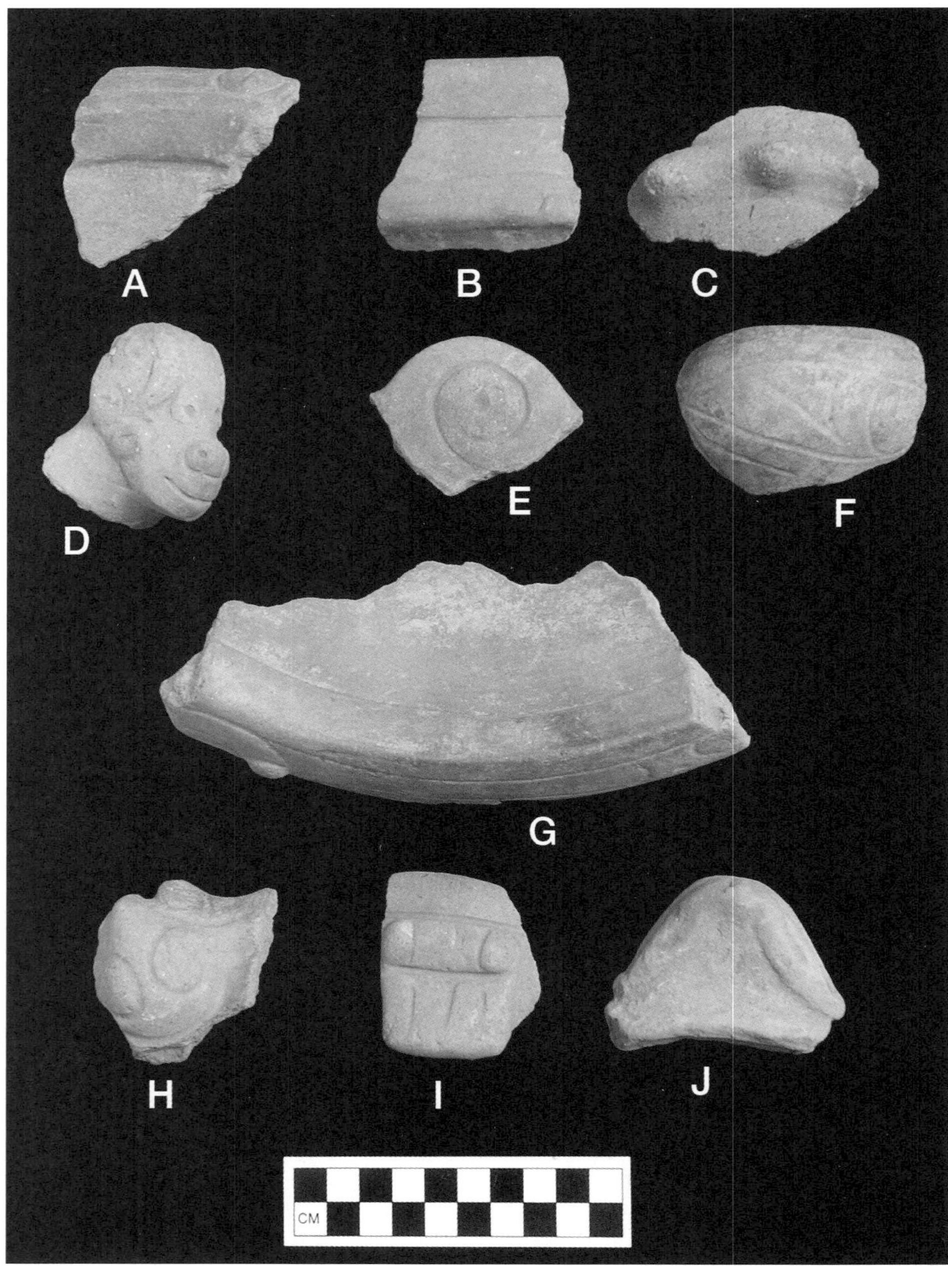

FIGURE 50. Incised and modeled-incised pottery from the Quinam site (Excavation 1). **A, B, D–J,** Erin complex. **C,** Bontour complex. *Location finds:* A, F–H, A2: 20–40 cm. B, C, A1: 0–20 cm. D, A1: 20–40 cm. E, A6: 20–40 cm. I, J, A4: 20–40 cm. *YPM Catalog Nos.:* (A) ANT 180357; (B) ANT 180154; (C) ANT 180201; (D) ANT 180298; (E) ANT 180598; (F) ANT 180387; (G) ANT 180357A; (H) ANT 180389; (I) ANT 180534; (J) ANT 180536.

the flaring open bowl of Form 3 (Figure 40B:3) gradually becomes less frequent. It is occasionally oval in horizontal cross section. Form 26 is unknown from the Palo Seco context. Concavo-convex flanged rims and interiorly flattened rims now outnumber direct rims, including interiorly or exteriorly thickened or beveled, tapering, and T-shaped forms, with or without flattened lips (Boomert 2000, table 11). Vessel Forms 4 (Figure 40B:4), 5 (Figure 40B:5), 8, 11, and 15 remain rare. Form 11 is not known from the Palo Seco context in Rouse's excavations in southwest Trinidad. Squarish to rectangular dishes of Form 8 and necked jars of Form 15 belong to the most conspicuous of these vessel classes. A flaring open bowl of Form 3, a Form 15 necked jar, a Form 6 flanged bowl, and a bottle of Forms 21/22 (Figure 41:8) have been encountered as mortuary gifts, indicating that these vessel shapes were used in domestic as well as ceremonial contexts. The frequency of use of two common Cedros vessel shapes, Forms 2 and 3, seems to decrease throughout the Palo Seco complex; the same applies to Form 4, which was rare in Cedros times.

The ceramic repertoire of the Palo Seco complex is much more varied than that of the Cedros period. Nine new vessel shapes—Trinidad and Tobago Saladoid Vessel Forms 7, 12, 16, 17, 18, 19, 20, 23, and 24—now make their appearance, of which Forms 17, 18, 23, and 24 are unknown from Rouse's excavations in southwest Trinidad (see Figure 38). They include two new bowl shapes: the small to medium-sized, oval to "hammock-shaped" bowl with dependent restricted orifice and composite contours of Form 7, and the small bowl with dependent restricted orifice, composite contours, and a triangular, occasionally hollow, rim of Form 12 (Figure 41:3). In the latter case, the vessel rim contains tiny clay pellets. Both Forms 7 and 12 resemble Barrancoid vessel shapes, notably Erin Forms 1, 4, and 7, indicating that they can be dated to the latter half of the Palo Seco complex (Boomert 2000:158).

Three rare jar forms are new. The medium-sized jar with independent-restricted orifice and composite contours of Form 16 (Figure 41:6) may be relatively late in the Palo Seco sequence. In contrast, the typically "pear-shaped" large jar with independent restricted orifice and complex contours of Form 23, has shows a slightly bent, convex neck decorated with an anthropozoomorphic face design and three vertical, D-shaped strap handles, probably dates from the transition between the Cedros and Palo Seco complexes. Finally, the double-spouted jar with independent restricted orifices and composite contours of Form 24 was apparently influenced by the typically Barrancoid double-spout-and-bridge bottle of Erin Form 12 and, consequently, may date from Late Palo Seco times.

Four special vessel shapes make their appearance in the Palo Seco complex: (1) the relatively small and rare asymmetrical bottle-like vessel with independent restricted orifice and oval horizontal cross section of Form 17; (2) the small to medium-sized, spouted vessel with independent restricted orifice and composite contours of Form 18; (3) the small "nostril" or "sniffing" bowl with restricted orifice, simple contours, and a pair of tube-like extensions of Form 19 (Figures 41:7, 62C); and (4) the small "double" vessel consisting of two interconnected bowls with

FIGURE 51. Painted, incised, and modeled-incised pottery from the Quinam site (Excavation 1). **A–F, I,** Erin complex. **G, H,** Palo Seco complex. *Location finds:* A, G, H, A1: 40–60 cm. B, A3: 40–60 cm. C, A2: 40–60 cm. D, F, A4: 40–60 cm. E, A6: 40–60 cm. I, A5: 40–60 cm. *YPM Catalog Nos.:* (A) ANT 180658; (B) ANT 180807; (C) ANT 180693; (D) ANT 180897; (E) ANT 180992; (F) ANT 180900; (G) ANT 180636; (H) ANT 180656; (I) ANT 180759.

unrestricted orifices and simple contours of Form 20. The spouted vessel of Form 18 clearly dates from the latter part of the Palo Seco complex.

These new vessel classes can be assigned to several functional categories. The Form 7 bowl was probably intended for the storage of dry goods or for food preparation without heating. Because it was used as a mortuary gift, this bowl shape may have served domestic as well as ceremonial purposes. The use of the Form 12 bowl as a mortuary gift and rattling device suggests that it was primarily a serving vessel in ceremonial contexts. The jars of Forms 14, 16, and 23 were most likely used to store liquids, possibly cassava beer, or as transport vessels. The four special vessel forms may have functioned as primarily ceremonial pottery. The asymmetrical bottle-like vessel of Form 17 and the spouted vessel of Form 18 perhaps served for storing liquids or for drinking by pouring. The double-spouted "nostril" bowl of Form 19 was clearly used by shamans for pouring tobacco or pepper juice into the nose to induce an ecstatic-visionary trance. Finally, the "double" bowl of Form 20 may have been a serving vessel in ceremonial contexts.

Base Forms. The four common Cedros complex base forms continue to be used, and the ring bases of Base B are especially typical. Three types appear for the first time: flat with convex upper side (E), flat and perforated (F), and footed (G). The latter two base forms are unknown from Rouse's excavations in southwest Trinidad. Base G resembles Base D of the Erin complex and, consequently, is dated to Late Palo Seco times.

Handles. Handles are predominantly vertical, peg-topped strap handles, D-shaped in cross section, just as during the Cedros complex, although horizontal handles, placed on top of vessel rims, occur as well (Figures 47C, D and 56F).

DECORATION

Painting, incision, punctation, simple modeling, and complex modeling form the five techniques of decoration, just as in Cedros times.

Painting. Pre-fired painting in red, white, and black, notably WOR motifs, remains typical (Figures 46J, 47A, B, D, and 56G, H). In addition, painting in yellow and pink occurs rarely. Direct rims as well as concavo-convex flanged rims are often painted red (Figure 46I, K). Polychrome red, white, and black painting is rare. Motifs which appear for the first time include black lines separating red- and black-painted areas, black-on-white painted designs, red- and white-filled incised lines, and plain lines excised in white-painted areas. Most likely Late Palo Seco motifs comprise red-and-black painted designs, reflecting Barrancoid influence. The same may apply to red painted motifs outlined by incised lines. Interestingly, pigment was often applied to accentuate certain plastic elements, such as the features of faces. Black-filled incised designs and red painted crosshatching have disappeared. The inner surfaces of several potsherds show a thick black coating, just as during Cedros times.

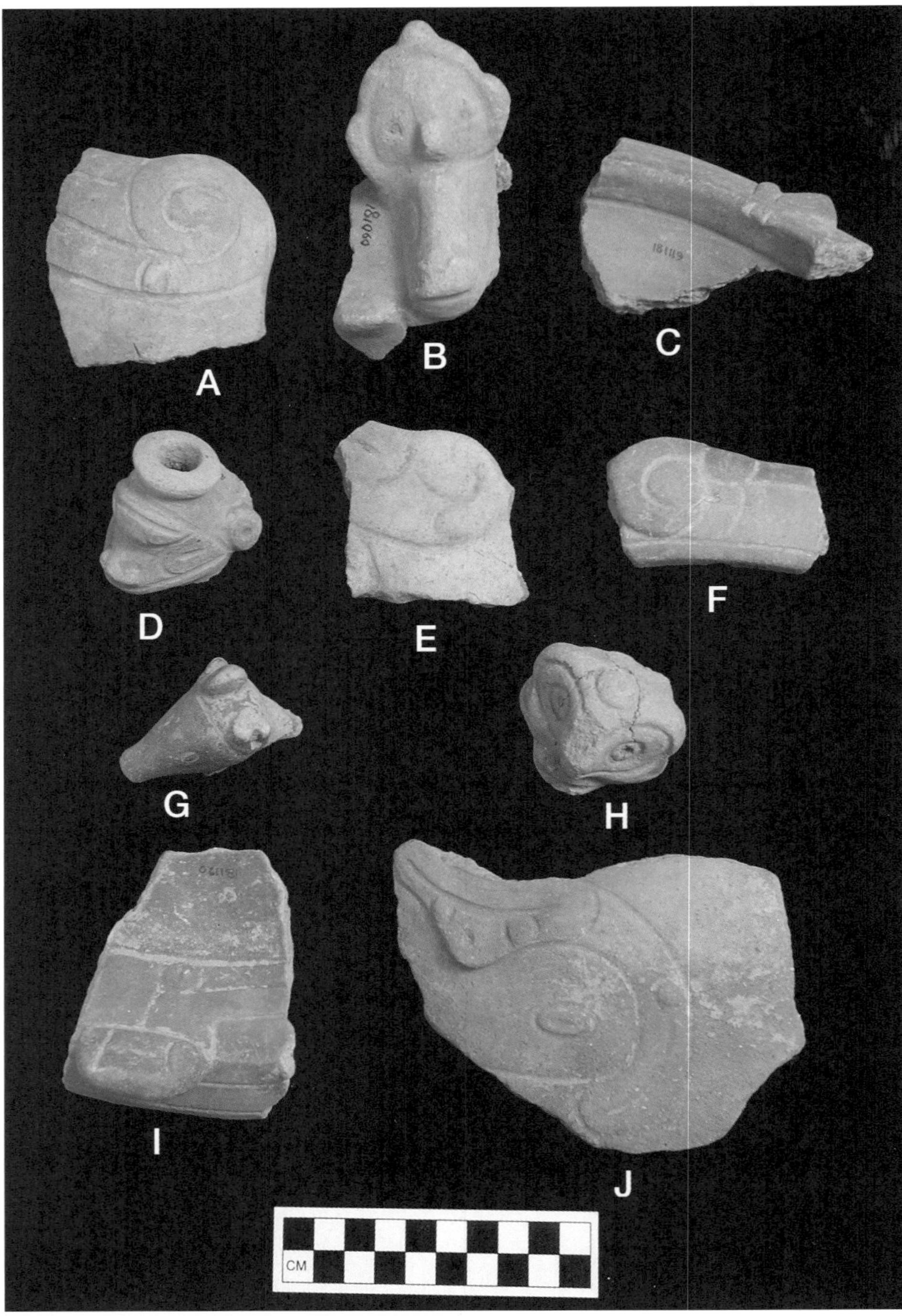

FIGURE 52. Incised and modeled-incised pottery from the Quinam site (Excavation 1). **A–F, H–J,** Erin complex. **G,** Palo Seco complex. *Location finds:* A, A3: 60–80 cm. B, A1: 60–80 cm. C, D, F–H, A2: 60–80 cm. E, A4: 60–80 cm. I, A1: 60–80 cm. J, A5: 60–80 cm. *YPM Catalog Nos.:* (A) ANT 181261; (B) ANT 181060; (C) ANT 181119; (D) ANT 181104; (E) ANT 181329; (F) ANT 181118; (G) ANT 181338; (H) ANT 181174; (I) ANT 181120; (J) ANT 181403.

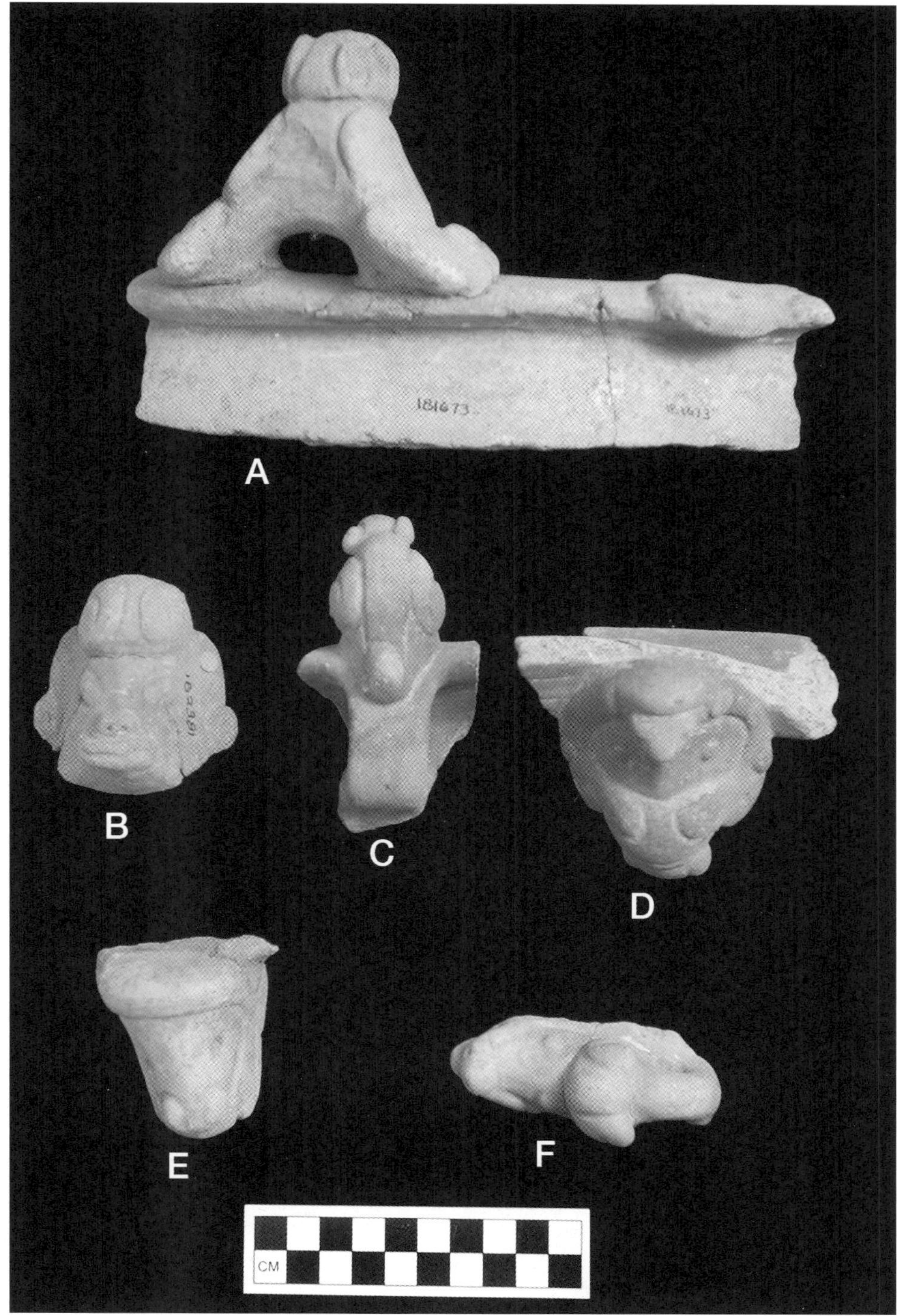

FIGURE 53. Modeled-incised pottery of the Erin complex from the Quinam site (Excavation 1). *Location finds:* A, E, A5: 80–100 cm. B, A4: 140–160 cm. C, A4: 100–120 cm. D, A3: 80–100 cm. F, A4: 80–100 cm. *YPM Catalog Nos.:* (A) ANT 181673; (B) ANT 182381; (C) ANT 181929; (D) ANT 181557; (E) ANT 181679; (F) ANT 181618.

Figure 54. Modeled-incised pottery of the Erin complex from the Quinam site (Excavation 1). *Location finds:* A, D–F, A3: 100–120 cm. B, A4: 80–100 cm. C, A6: 80–100 cm. G, A4: 100–120 cm. *YPM Catalog Nos.:* (A) ANT 181795; (B) ANT 181617; (C) ANT 181643; (D) ANT 181771; (E) ANT 181798; (F) ANT 181841; (G) ANT 181922.

Incision. ZIC motifs are much less frequent than during Cedros times, but certainly not absent as Rouse (1947) holds (Figure 49C). ZIC seems to have evolved gradually into a specific type of thin-line incised, or rather engraved, design (Faber-Morse 2010) of single or multiple parallel lines and semi-circles, spirals, and stepped lines. Interestingly, these lines seem to disappear first at the Palo Seco sites of southwest Trinidad. Elsewhere, especially in the northern and eastern parts of the island, the designs remained in use, albeit infrequently, until as late as the end of the Palo Seco complex. Broadline incised curvilinear and rectilinear motifs continue (e.g., Figure 56A, E). The latter often adorn the top of flanged or triangular rims of bowls of Forms 2 and 6 in the form of a few parallel lines or of "friezes" consisting of such lines, alternating horizontally and vertically (Figure 51G). Both the general emphasis on broadline incision and the specific motifs dominating throughout the Palo Seco complex may reflect Barrancoid influence.

Punctation. This is extremely rare, just as in Cedros times (e.g., Figure 51H).

Simple modeling. There is no change in technique of execution and motifs compared to the Cedros complex. Common motifs such as buttons-and-bars, applied to vessel rims or bellies, and triangular side lugs surmounting D-shaped strap handles continue (Figures 46I, 48C, H, J, K, 49B, E, 51H, 55G, and 62G, K). Some clearly represent Barrancoid imitations (Figure 46H). Four-lobed and so-called "dimpled" vessel rims appear for the first time.

Complex modeling. Modeled-incised, geometric, and anthropozoomorphic head lugs become much more varied and complicated during Palo Seco times (Figures 46B, C, F, G, L, 47A–E, G–I, 48F, G, 49A, D, 52G, 55A–F, 56D, F, 60E, 61A, B, E–H, and 62D, F, H, I). Such lugs adorn primarily the rims of bowls, handles, bottle spouts, and stoppers. Small modeled-incised geometric *adornos*, showing influence from Barrancoid prototypes, are found especially on vessel walls and rims. Typically Saladoid hollow, mammiform lugs continue; trapezoidal side lugs are new. Both become rare toward the end of the Palo Seco sequence. Characteristic elements include eyes indicated by unmodified pellets, punctations, punctated pellets, short gashes or slits (e.g., Figures 46C, D, F, G, L, 47A, C–E, G–I, 52G, 55A–F, and 62H–J). Head lugs consisting just of a few punctated nubbins are rare now. Concave backs remain common. A new feature is the placement of the nose above the eyes of the animal or human-like being represented (Figure 46G, L). Hollow, biomorphic *adornos* containing a series of clay pellets or stone pebbles, which produce a rattling sound when moved quickly, similarly appear for the first time (e.g., Figure 56F). Human- and animal-like head lugs showing Barrancoid influence are much less stylized than their Erin counterparts.

Although indeed the variability in execution is wide, a few major themes and design elements can be identified. Most of the head lugs are cylindrical or spherical and solid or hollow. Eyes formed of three long, slanting incisions seem to be typical of the latter part of the Palo Seco sequence. Various biomorphic head lugs show

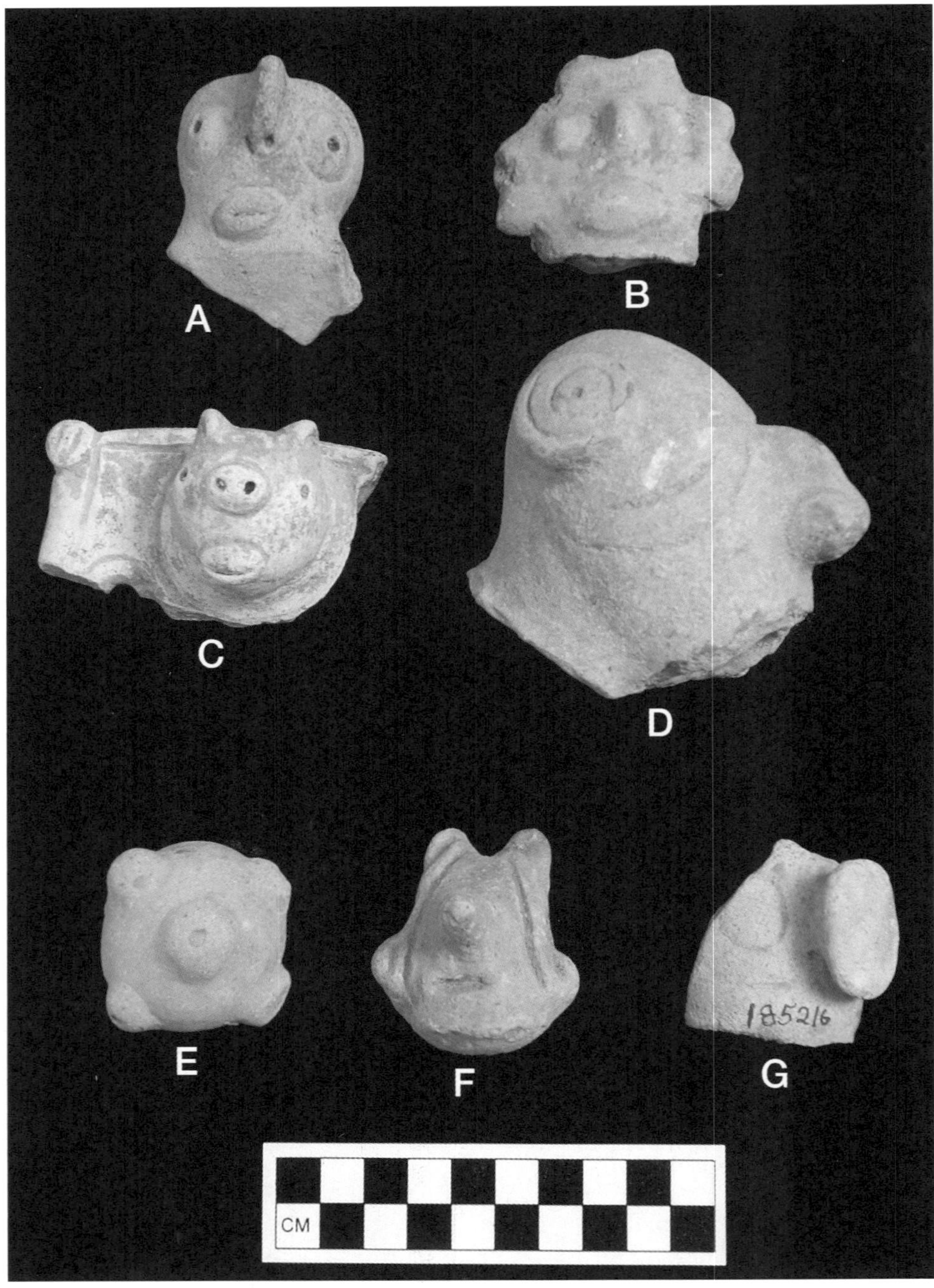

FIGURE 55. Modeled and modeled-incised pottery of the Palo Seco complex from the Quinam site (Excavation 2). *Location finds:* A, B, G1: 20–40 cm. C, G3: 60–80 cm. D, G6: 60–80 cm. E, G3: 80–100 cm. F, G5: 80–100 cm. G, G5: 100–120 cm. *YPM Catalog Nos.:* (A) ANT 182108; (B) ANT 182709; (C) ANT 183833; (D) ANT 184208; (E) ANT 184296; (F) ANT 184956; (G) ANT 185216.

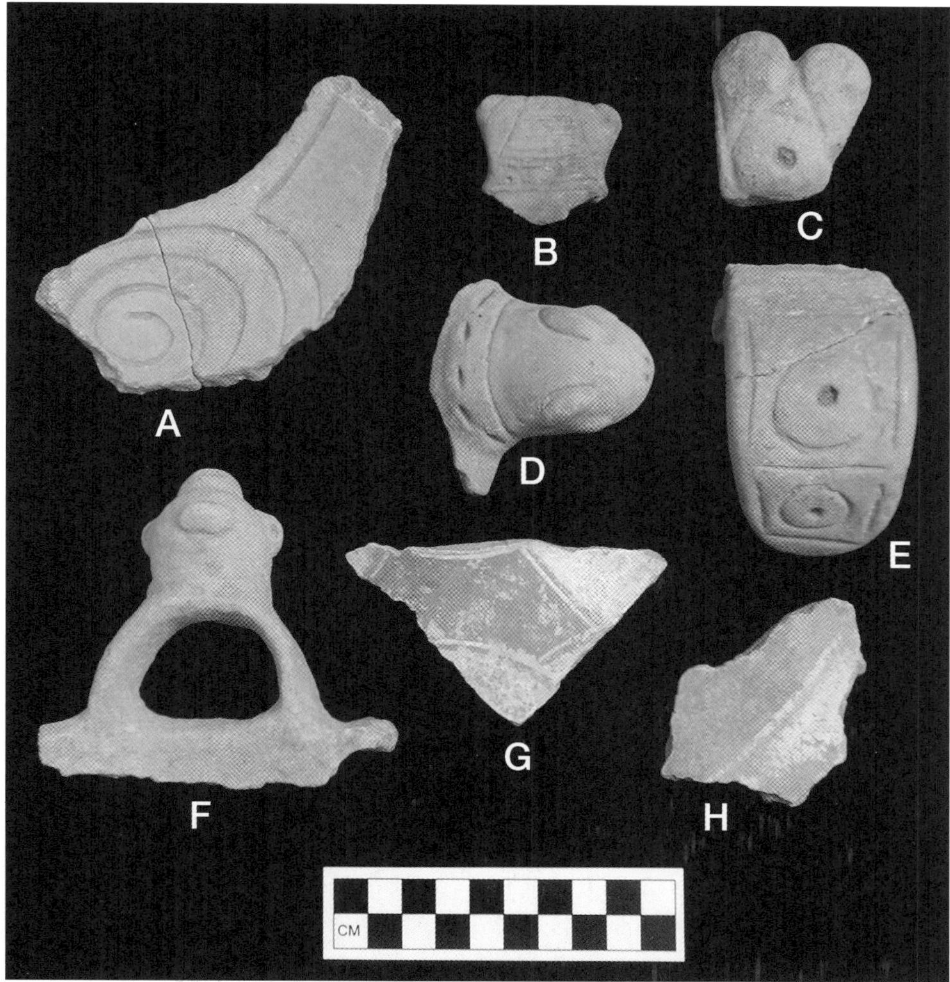

FIGURE 56. Painted, incised, and modeled-incised pottery from the Quinam site (Excavation 3). **A–B, D–H,** Palo Seco complex. **C,** Erin complex. *Location finds:* A, F, M1: 40–60 cm. B, C, M1: 20–40 cm. D, E, G, H, M1: 60–80 cm. *YPM Catalog Nos.:* (A) ANT 185439/185440; (B) ANT 185405; (C) ANT 185611; (D) ANT 185403; (E) ANT 185608; (F) ANT 185525; (G) ANT 185536; (H) ANT 185535.

Erin-influenced modeled-incised noses provided with nostrils, typically indicated by short spirals encircling a single punctation (Figure 46F). Other themes shown by both human- and animal-like *adornos* include protruding foreheads, "topknots" (distinctive, round pompon-like tops), and semi-circular, often multipointed, head-dresses (e.g., Figure 55D). The animal species portrayed by the zoomorphic head lugs remain difficult to identify, but include monkeys, bats, peccaries, armadillos, felines, birds, possibly turtles, frogs or toads, unidentified reptiles, and caimans. The bird-like *adornos* represent king vultures (resembling those of the Erin complex), pelicans, and parrots or macaws (Figures 47B, 49D, 62J). Circular, disc-like lugs, typically showing incised spirals, often encircling a single punctation, probably

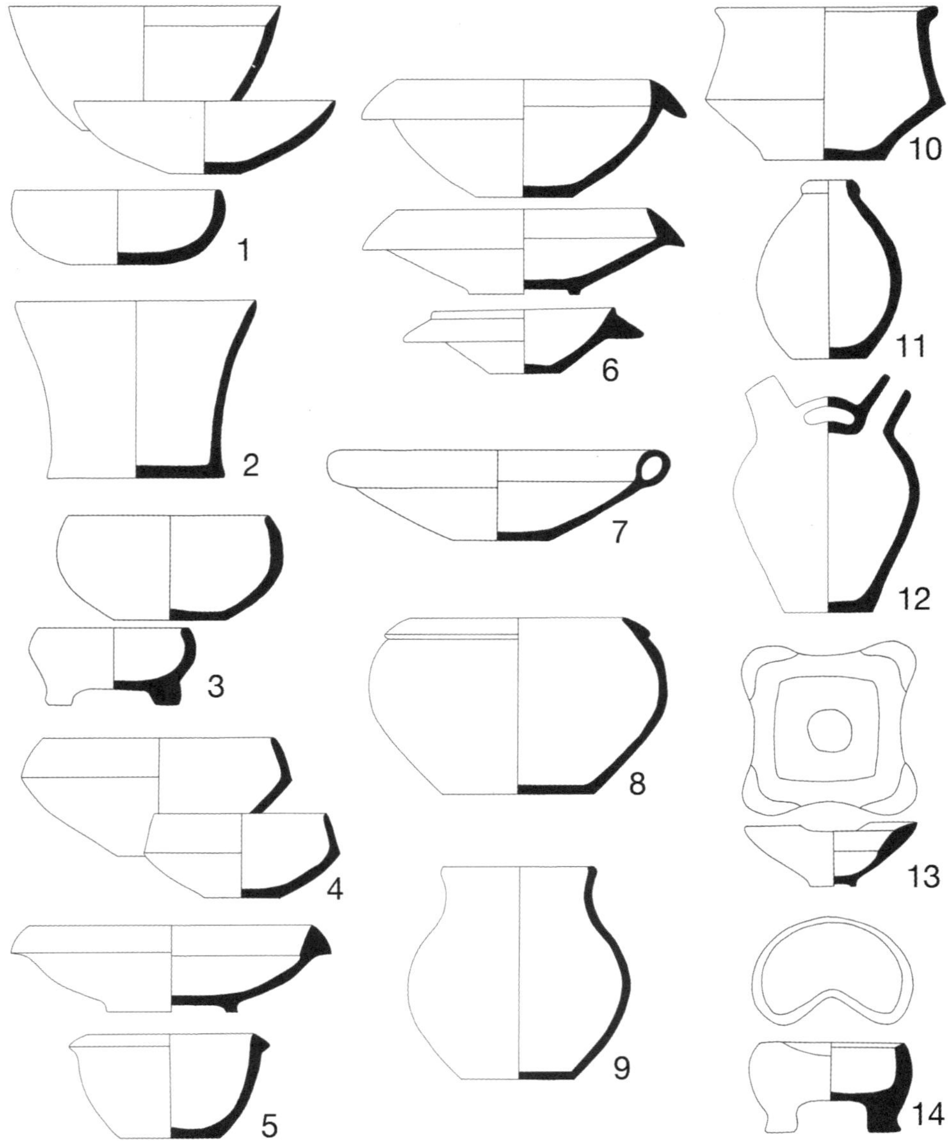

Figure 57. Reconstructed Erin complex vessel forms from Trinidad and Tobago (not to scale). Adapted from Boomert (2000, figs. 48 and 49).

represent conventionalized bird heads. The importance of effigy vessels in the Erin complex is duplicated by that in Palo Seco. Indeed, many of the head lugs discussed derive from vessels representing an entire animal.

Other Kinds of Clay Artifacts

Three categories of ceramic artifacts are known from the Palo Seco complex:

griddles, pottery cylinders, and spindle whorls. The griddles resemble those from Cedros times, with flat unmodified or thickened upturned rims. Several fragments of pottery cylinders were encountered associated with Palo Seco ceramics at the Palo Seco site. Finally, Rouse's Excavation 1 at Erin yielded a piece of a possible spindle whorl.

Erin (Erinan, Barrancoid)

Undoubtedly, Erin ceramics represent both the artistic and technological climax of pre-Columbian culture in Trinidad, or for that matter the entire Caribbean. Pottery of the Erin complex has been found at sixteen sites in all in central and southern Trinidad, but only in substantial amounts exclusively in the south and southwest of the island, notably at the Chagonary, Erin, Quinam, and La Lune 1 multicomponent settlement sites. Elsewhere, such as at Palo Seco and Manzanilla 1, it is restricted to "trade" pottery in the Palo Seco complex context (Boomert 2000:202–204). Erin ceramics are consistently encountered in close association with Palo Seco pottery, suggesting largely simultaneous manufacture and use of both wares in Trinidad. Rouse's excavations at Erin, the type site, yielded Erin complex ceramics in the upper and middle portions of his Excavation 1, levels 1 to 8 (0 to 160 cm), and the top part, levels 1 and 2 (0 to 40 cm), of Excavation 2. In addition, Erin pottery was encountered throughout Excavations 1 and 2 and in the top portion, levels 1 to 4 (0 to 80 cm), of Quinam, Excavation 3, associated with Palo Seco ceramics, and in the upper part, level 1 (0 to 20), of Quinam, Excavation 4, accompanied by Bontour pottery. Palo Seco, Excavations 1 and 2, similarly yielded Erin pottery only in the upper parts of the midden deposits, levels 1 to 4 (0 to 80 cm) and levels 1 and 2 (0 to 40 cm), respectively, associated with Palo Seco ceramics.

Material

Erin pottery reflects the highest standard of technological skills of all prehistoric cultural complexes of Trinidad. The vessels were made by coiling and were fired in an open fire. If recognizable, the coil width is 1.7 to 1.8 cm. Vessel wall thickness varies from 6 to 14 mm, on average 8 to 10 mm. Surfaces are generally well smoothed or burnished, showing careful and even treatment. The pottery is relatively hard, 4 to 4.5 on the Mohs scale. Oxidization is incomplete. Sherds are yellowish gray to gray on the inside and outside surfaces and dark gray in cross section. A dense, gritty paste is typical. Temper consists predominantly of abundant quantities of fine to medium-sized angular particles of apparently deliberately crushed, mainly transparent quartz. Temper is visually some 40% of the paste. Many temper particles stick out from the surface, resulting in an unpleasant touch. Few sherds are tempered heavily with fine quartz sand, whereas pieces showing a combination of crushed quartz or fine sand and *cauixí* (freshwater sponge spicules), or exclusively *cauixí* temper, are rare. Most of the fine sand-tempered sherds, as well as those containing *cauixí*, seem to date from the end of the Erin complex. These Late Erin pieces are typically less thick (generally about 6 to 7 mm) than the ones containing

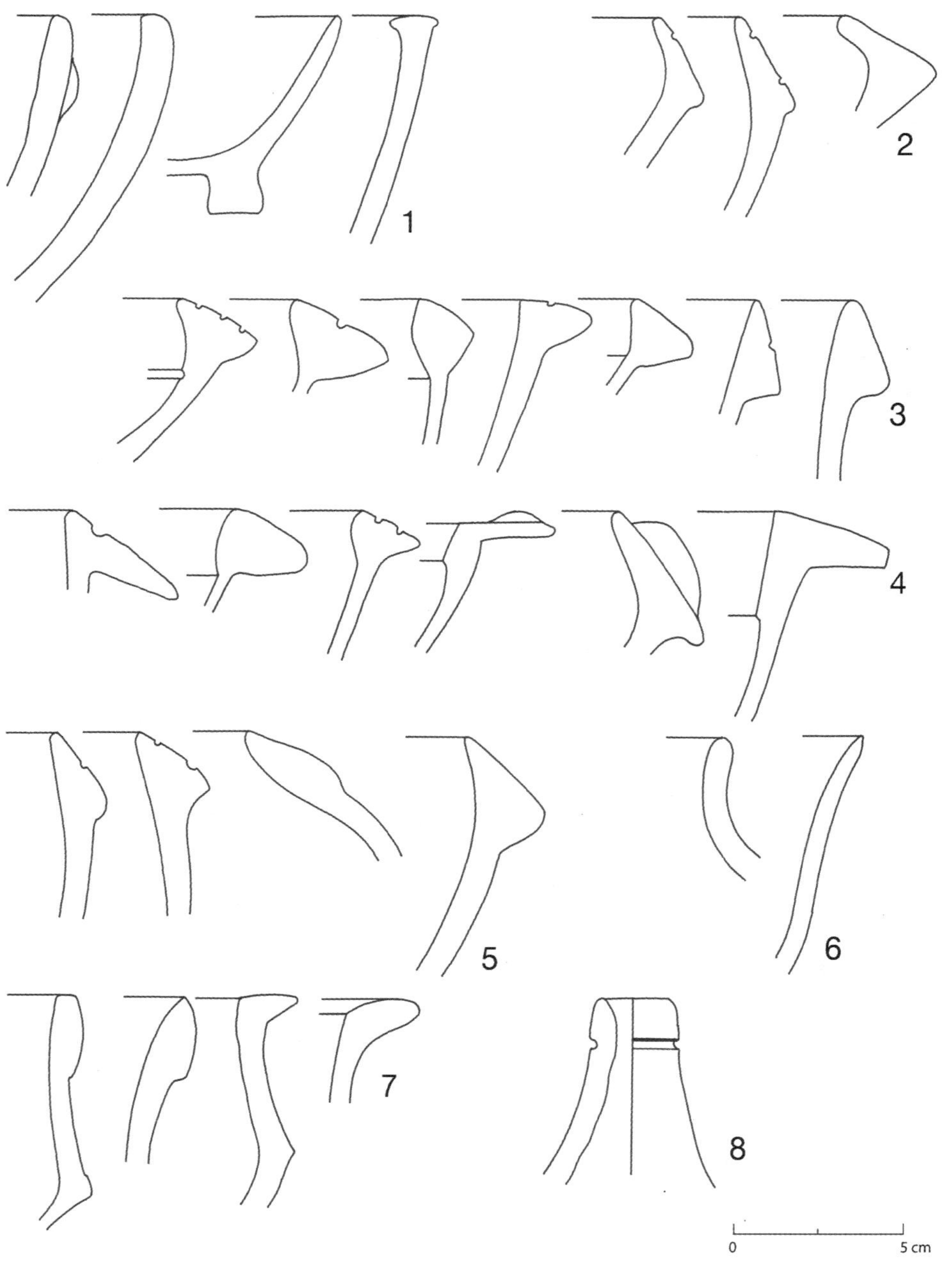

Figure 58. Erin complex vessel forms and rim types from the Quinam (Excavations 1 and 2) and Erin (Excavation 1) sites. (1) Erin Form 1; (2) Erin Form 4; (3) Erin Form 5; (4) Erin Form 6; (5) Erin Form 8; (6) Erin Form 9; (7) Erin Form 10; (8) Erin Form 11.

particles of deliberately crushed quartz, which typify the ceramics of the Erin complex from its onset. In addition, griddles are often tempered with *cauixí*. Finally, the upper portion of Rouse's Excavation 1 at Erin yielded an undecorated potsherd tempered with micaceous sand. This may represent a "trade" piece from the Lower Orinoco Valley (see Boomert 2000:119).

SHAPE

Vessel Shapes. In all, there are fourteen distinguishable vessel shapes (Figure 57). They show typically small orifice diameters; only Forms 4 and 10 reach medium-sized proportions. Externally thickened rims, triangular in cross section, and heavy flanges, occasionally flattened on the inside, are typical. Direct rims frequently have flattened lips. Hollow rims are rare. Three vessel shapes, Forms 1, 5, and 7, are bowls with unrestricted orifices and simple contours. Form 1 is a round or oval, even "hammock-shaped," bowl or basin with direct or interiorly thickened rim (Figures 50F, 58:1). Form 5 may be round or oval in horizontal cross section as well. It shows a heavy, exteriorly thickened rim, triangular in cross section (Figures 50G, 54A, and 58:3). Form 7 is a rare vessel shape with an outward-bulging hollow rim. Two bowl or jar shapes, Forms 3 and 8, show restricted orifices and simple contours. Form 3 has a direct, unmodified rim, whereas Form 8 has an exteriorly thickened rim, triangular in cross section (Figure 58:5). Two bowl shapes, Forms 4 and 6, have dependent restricted orifices and composite contours. The "biconical" bowl of Form 4 has a direct, unmodified rim (Figure 58:2), while Form 6, which may be round, oval or even "hammock-shaped" in horizontal cross section, shows a heavy, occasionally almost T-shaped, flanged rim (Figures 50A, 51E, and 58:4). Several of these bowl shapes, notably Forms 3 and 8, may have been used for the storage of dry goods or for food preparation without heating, while Forms 1, 4, 5, 6, and 7 may have functioned as vessels for serving, drying, or displaying food.

Vessels probably used for cooking include Form 9, a jar with an independent restricted orifice, inflected contours, and a direct unmodified rim (Figure 58:6), and Form 10, a jar with restricted orifice and composite contours showing a slightly concave profile above its corner point and exteriorly thickened rims, triangular in cross section (Figures 50B, 58:7). Two vessel shapes, Forms 11 and 12, clearly functioned as storage for liquids or as traveling canteens. Form 11 is a jar or bottle with independent restricted orifice and exteriorly thickened rim and composite contours (Figure 58:8). The double-spout-and-bridge-bottle of Form 12 has an independent restricted double orifice, an exteriorly thickened rim, and composite contours (Figures 47F, 52D). Finally, three rare vessel shapes, Forms 2, 13, and 14, may have been used for serving, drying, or displaying food. Form 2 is a jar with unrestricted orifice, simple contours, and direct unmodified rims. Form 13 is a squarish to rectangular bowl with unrestricted orifice and an exteriorly thickened rim. Form 14 is a footed bowl with unrestricted orifice, kidney-shaped in horizontal cross section, with nearly vertical, simple contours. Forms 2, 7, 13, and 14 are not known from Rouse's excavations in southwest Trinidad.

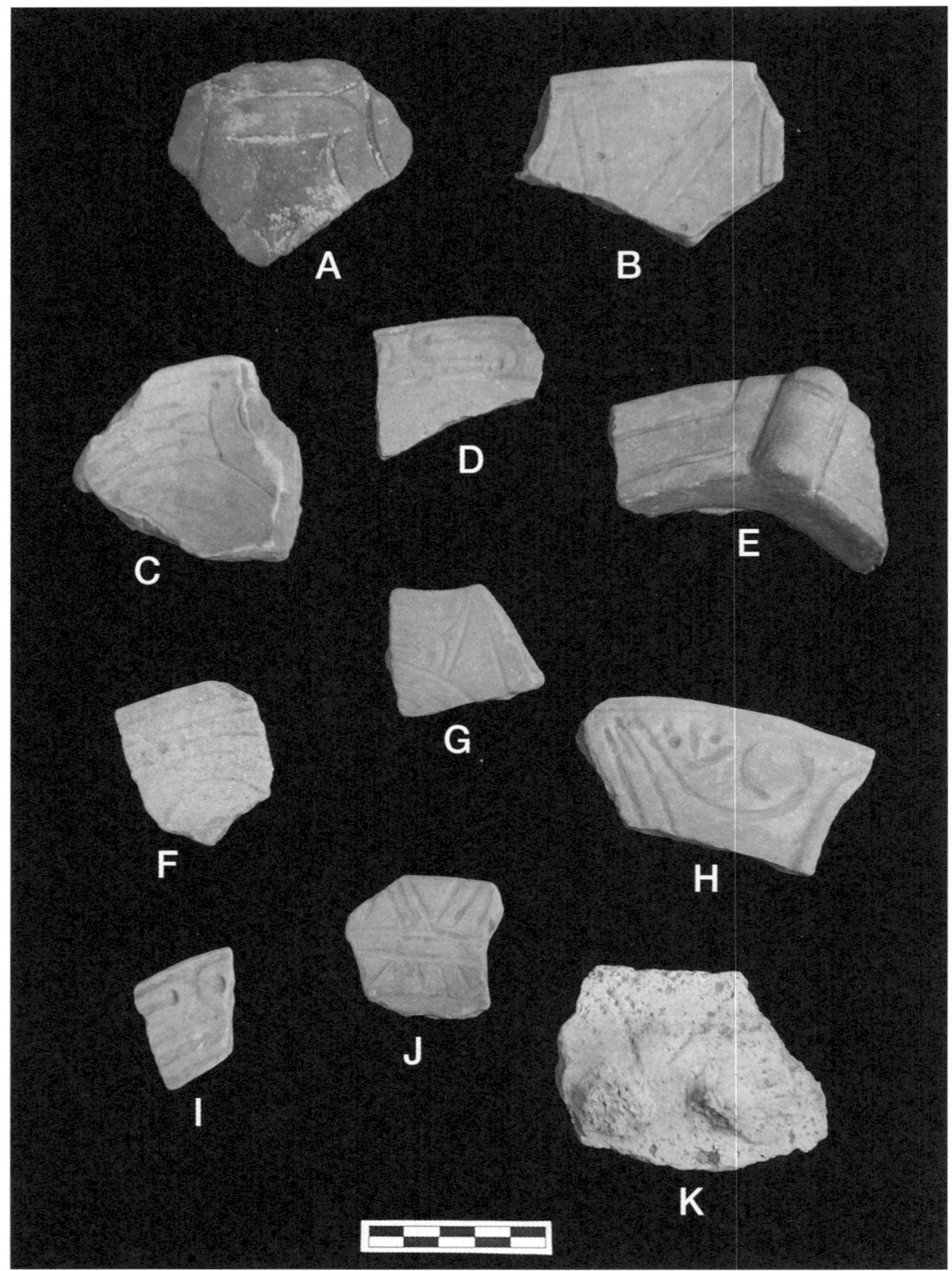

FIGURE 59. Incised, punctated, and modeled-incised pottery from the Erin site (Excavation 1). **A–J,** Erin complex. **K,** Bontour complex, *Location finds:* A, B2: 20–40 cm. B, D, B1: 40–60 cm. C, B4: 40–60 cm. E, B3: 20–40 cm. F, B4: 60–80 cm. G, B4: 20–40 cm. H, B2: 40–60 cm. I, B2: 60–80 cm. J, C4: 40–60 cm. K, B3: 40–60 cm. *YPM Catalog Nos.:* (A) ANT 169296; (B) ANT 169925; (C) ANT 170292; (D) ANT 169934; (E) ANT 169387; (F) ANT 170840; (G) ANT 169607; (H) ANT 170052; (I) ANT 170702; (J) ANT 170326; (K) ANT 170130.

Base Forms. There are four base forms: flat (A), annular and outflaring or "ring shaped" (B), rounded (C), and footed (D). Bases A and B show unmodified basal angles. Base D is provided with three or four feet, which are squarish, circular, or oval in cross section.

Handles. Handles are predominantly vertical, D-shaped strap handles often peg-topped or surmounted by zoomorphic head lugs (Figure 53C). Horizontal handles, placed on top of vessel rims, are rare (Figure 53A).

DECORATION

Five techniques of decoration can be distinguished in the Erin complex: painting (including smudging and polishing), incision, punctation, simple modeling, and complex modeling. Note that the incised, punctated, and modeled-incised decorative motifs associated with the Erin ceramics, which are tempered with fine quartz sand, *cauixí*, and a combination of both and which, consequently, date from the final part of the Erin cultural continuum, display a definite loosening of the strict standards of decoration shown by the "classic" pottery of the Erin complex, which is tempered predominantly with dense amounts of crushed quartz particles. Undoubtedly, this "degenerate version" of Erin ceramics reflects a gradual disappearance from the potters' minds of the spiritual templates previously determining vessel manufacture and decoration during Late Erin times.

Painting. Pre-fired painting in red is common; white and black painting (or smudging) is rare. All-over monochrome (dark) red painting of exterior vessel surfaces and that of the labial flanged rims of vessels predominate. Incised lines filled with red pigment are rare. Black smudging and subsequent polishing is infrequently found, notably at the Erin and Quinam sites. Red painted designs may be combined with black-smudged and polished areas, separated by incised lines, whereas incised lines are sometimes filled with white paint.

Incision. This is a typically Erin decorative technique; it occurs especially in association with modeling, red painting, or both. Incised lines are invariably wide and U-shaped, measuring about 1.5 to 3 mm across. They are occasionally polished. Motifs comprise straight parallel lines, found especially on flanged rims, curved lines, circles, spirals, volutes, lines ending in crossbars, lines ending in dots, Y-shaped lines, Y-shaped lines ending in dots, "telephone clamp" designs (straight or curved lines turning into volutes at both ends), "feather" motifs (Y-shaped lines of which the arms are connected by a short line, forming a triangular area at the base of the bifurcation that may or may not be "excised" entirely), and, finally, short oblique lines arranged as zigzags (Figures 46A, 50A, B, E, F, 51D, I, 52A, E, F, J, 54A, C, and 60G). Typically Late Erin potsherds show a quite loose, indeed sometimes distinctly haphazard and sloppy, distribution of incised design elements (Figures 59B–D, F–J, 66G, H).

Punctation. This is less common than incision and modeling. Motifs include fields of punctations bounded by incised lines and single punctations enclosed by incised rectangles, semicircles, and triangles. Late Erin punctation reflects a comparable disappearance of standards as with the incised designs (Figures 59H, 66G, H).

Simple modeling. This encompasses a series of modeled and modeled-incised designs, including circular to oval, punctated nubbins (*mamelones*), often surmounting D-shaped strap handles, small circular or rectangular appendages or side lugs, lobed or "dimpled" vessel rims, and, finally, incised-punctated wall bosses (Figures 50A, F, 51B, C, E, I, 52A, C, I, J, 53D, 54A, 59E, 60G, and 66G).

Complex modeling. This includes a wide variety of intricately modeled-incised, strongly sculptural, geometric and anthropozoomorphic head lugs. Geometric *adornos* are especially found on vessel walls, heavy rim flanges, or externally thickened rims. They include small, tabular lugs, spirals or volutes combined with modeled elements, and modeled-incised "telephone clamp" designs. Human- and animal-like head lugs surmount vertical strap handles and adorn vessel walls or bottle spouts (Figures 47F, 50G, 52D, 53A, B, D–F, 54F, G, 60B–D, and 61D). Furthermore, highly stylized modeled-incised human and animal limbs, hands, feet, bird claws, and tails are characteristic design elements (Figures 46A, 50I, 54E, 60A, H). The zoomorphic head lugs represent birds, including king vultures, parrots, and macaws (Figures 51A, 60F, 61C), and animals such as monkeys, possibly felines, bats, armadillos, snakes, caimans, and turtles. Some animal- and human-like head lugs show extruding tongues, jaws with exposed teeth, or both. Eyes consist of pits, pitted or unpitted buttons, incised ovals, doughnuts, or slits. Occasionally, zoomorphic head lugs have features derived from different animals rather than from one particular species (Figures 46E, 50J, 51F, 52B, and 53C). Head lugs showing vultures surmounting anthropomorphic heads, so-called "alter ego" motifs, probably represent shamans in ecstatic-visionary trance. Another type of *adorno* that combines anthropomorphic and zoomorphic features shows a long-snouted creature, often with a semicircular headdress. Four-lobed square vessels of Form 13 are especially associated with a special form of anthropomorphic head lug, coined the "Praying Man" theme by Harris (1978). The importance of modeled-incised design elements such as human and animal limbs, hands, feet, and such suggests that effigy vessels take a prominent position in the Erin vessel repertoire. Sloppy execution characterizes the limited number of modeled-incised head lugs dating from Late Erin times (Figures 50D and 66I).

Other Kinds of Clay Artifacts

Two types of ceramic artifacts are known from the Erin complex: griddles and pottery cylinders. The griddles resemble those of the Saladoid series. Intricately decorated, circular and most likely ceremonial, cylinders are not uncommon.

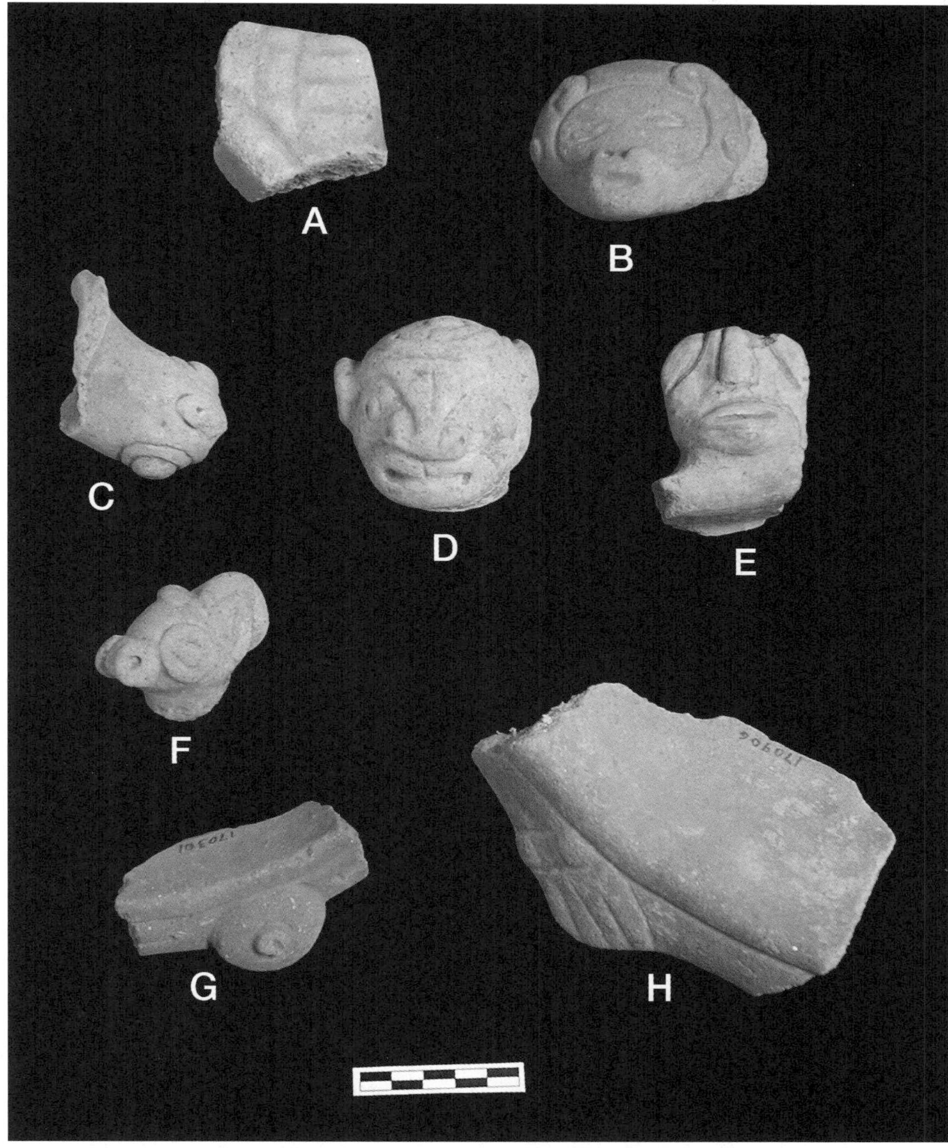

FIGURE 60. Incised, punctated, and modeled-incised pottery from the Erin site (Excavation 1). **A–D, F–H,** Erin complex. **E,** Palo Seco complex. *Location finds:* A, B, H, C4: 60–80 cm. C, A4: 40–60 cm. D, B2: 80–100 cm. E, B3: 0–20 cm. F, B2: 40–60 cm. G, C4: 40–60 cm. *YPM Catalog Nos.:* (A) ANT 170962; (B) ANT 170965; (C) ANT 170177; (D) ANT 171179; (E) ANT 169031; (F) ANT 170088; (G) ANT 170301; (H) ANT 170906.

Bontour (Guayabitan, Arauquinoid)

Pottery of the Bontour complex has been recovered from fifty-six sites in Trinidad, including predominantly settlement sites (midden deposits), less camp and bivouac sites (pottery deposits), and special activity sites (individual finds). A few Bontour midden deposits have yielded human burials (Boomert 1985,

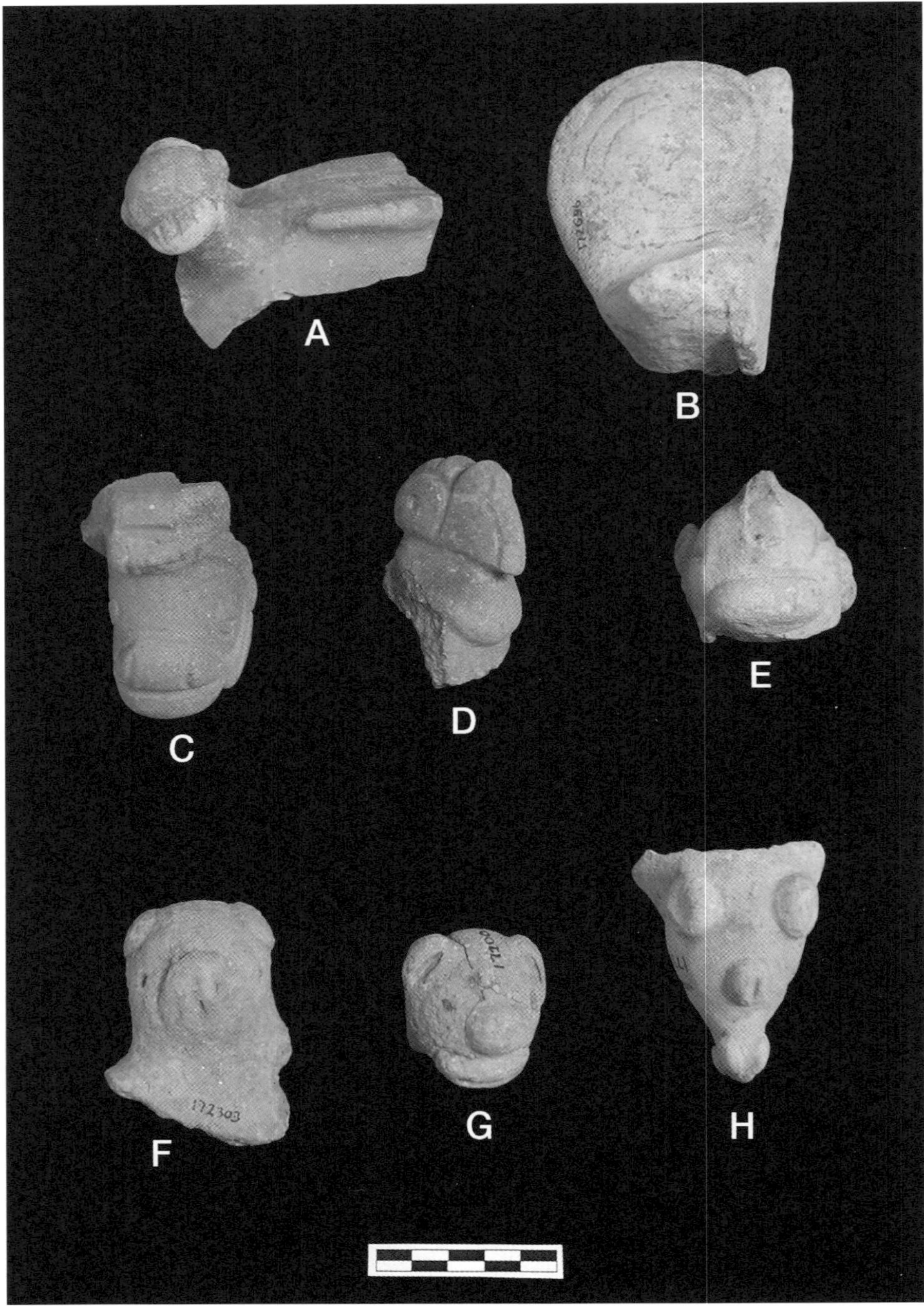

FIGURE 61. Modeled and modeled-incised pottery from the Erin site (Excavation 1). **A, B, E–H,** Palo Seco complex. **C, D,** Erin complex. *Location finds:* A, G, B1: 100–120 cm. B, D, B1: 120–140 cm. C, B2: 100–120 cm. E, B3: 120–140 cm. F, A4: 100–120 cm. H, C4: 140–160 cm. *YPM Catalog Nos.:* (A) ANT 172007; (B) ANT 172696; (C) ANT 172150; (D) ANT 172699; (E) ANT 172210; (F) ANT 172303; (G) ANT 172004; (H) ANT 173165.

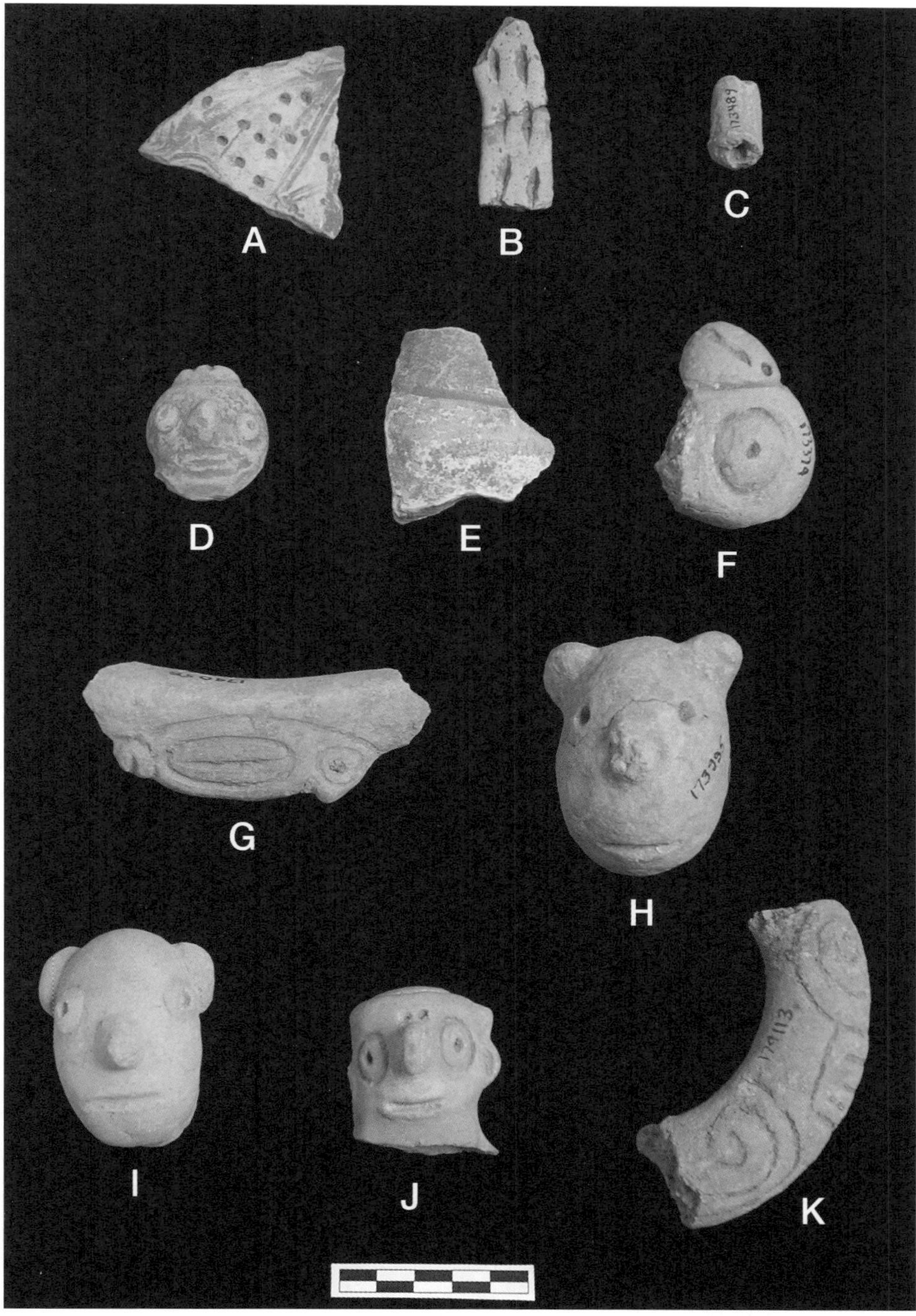

Figure 62. Painted, incised, punctated, modeled, and modeled-incised pottery from the Erin site (Excavation 2). **A, B,** Bontour complex. **C–K,** Palo Seco complex. *Location finds:* A, B, S2: 20–40 cm. C, Y2: 20–40 cm. D, S1: 60–80 cm. E, S1: 40–60 cm. F, S1: 0–20 cm. G, I–K, S2: 100–120 cm. H, S2: 80–100 cm. *YPM Catalog Nos.:* (A) ANT 173540; (B) ANT 173549; (C) ANT 173489; (D) ANT 173829; (E) ANT 173659; (F) ANT 173374; (G) ANT 174058; (H) ANT 173995; (I) ANT 174105; (J) ANT 174106; (K) ANT 174113.

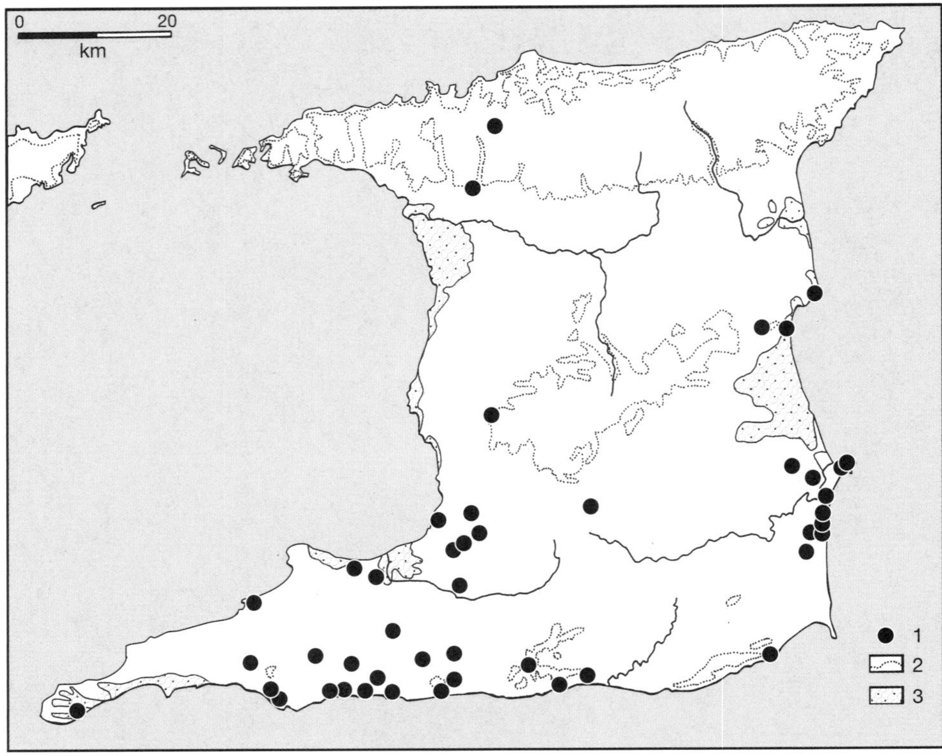

FIGURE 63. The location of Bontour complex sites on Trinidad. *Legend:* (1) Bontour sites; (2) 100 m contour; (3) swamps and marshes.

2000:495–510; Harris 1985; our current understanding of the Bontour complex includes the formerly separately distinguished St. Joseph and Marac assemblages [Boomert 1985]). Sites are distributed throughout the island, except for the northern littoral, but most sites are known from southern Trinidad (Figure 63). This is at least partially due to the history of archaeological investigation in Trinidad, which has seen a concentration of research in the south and central parts of the island. Bontour ceramics typify Rouse's Excavation 2 at the St. Joseph 2 site in the Northern Range. Bontour, the type site, and Excavation 4 at Quinam yielded Bontour pottery accompanied by small quantities of Late Erin ceramics, whereas a few Bontour "trade" pieces were encountered associated with Palo Seco and Late Erin earthenware in the upper portion, levels 1 to 3 (0 to 60 cm), of Rouse's Excavation 1 at Quinam. Sparse amounts of Bontour "trade" sherds are known also from the Palo Seco and Erin sites, associated with Palo Seco and Late Erin ceramics. The Palo Seco site yielded Bontour pottery in the upper portion, levels 1 to 4 (0 to 80 cm), of Excavation 1 and the top section, levels 1 to 2 (0 to 40 cm), of Excavation 2, whereas Bontour ceramics were found at Erin in the upper part, levels 1 to 5 (0 to 100 cm), of Rouse's Excavation 1 and the top section, levels 1 to 2 (0 to 40 cm), of his Excavation 2.

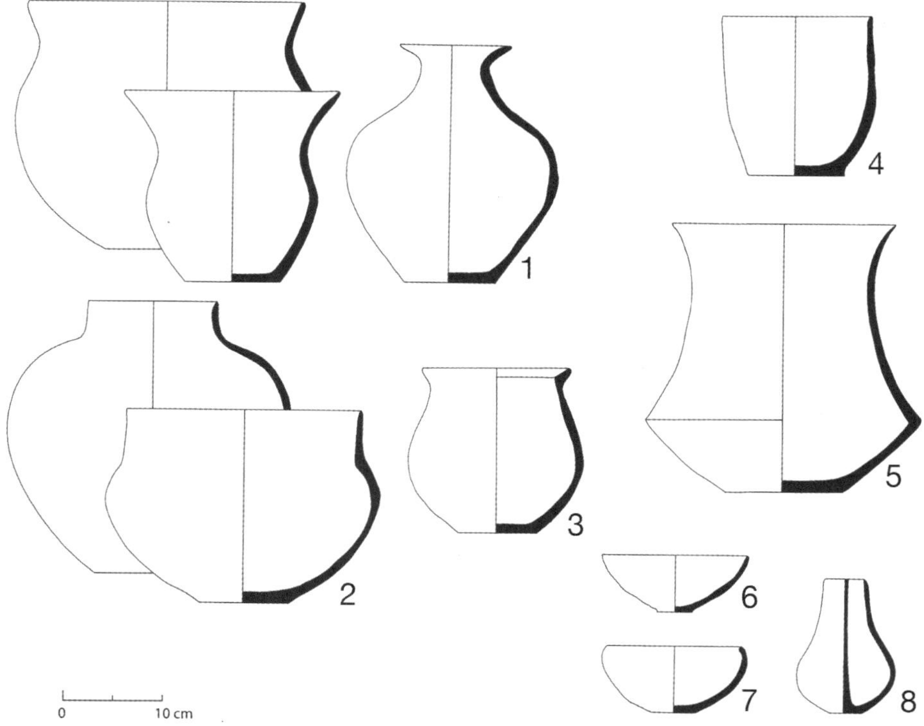

FIGURE 64. Reconstructed Bontour complex vessel forms from Trinidad (all to the same scale).

MATERIAL

Bontour pottery shows a considerable lessening in ceramic standards if compared to Cedros, Palo Seco and, especially, Erin earthenware. Vessels were made by coiling and were fired in an open fire. Wall thickness ranges from 6 to 10 mm, on an average 7 to 8 mm. Surfaces are generally poorly smoothed, showing careless treatment. The pottery is relatively soft, 2 to 2.5 on the Mohs scale. Oxidization is incomplete. Sherds are predominantly gray to tan or yellowish gray on the inside and outside surfaces and dark gray in cross section. At the Bontour sites of south and central Trinidad temper consists predominantly of abundant quantities of fine to medium-sized fragments of pounded shell (see Boomert 1985, table 1). Many sherds are pockmarked from the leaching of the particles of shell used as tempering material, adding to the drab appearance of Bontour ceramics. Temper is visually some 30% of the paste. Shell temper is preponderant at the Bontour site (93.7%) and at Quinam, Excavation 4 (40.6%). In north Trinidad, notably at St. Joseph 2, predominantly local river sand, characterized by medium-sized to coarse particles of quartz, feldspar, and micaschist, was used for tempering (76.5%). Here shell temper is less common (9.5%). Potsherds tempered with dense quantities of fine quartz sand are a minority ware, especially at Quinam, Excavation 4 (29.6%). This applies to *cauixí* temper as well (15.4%). *Cauixí* occurs in negligible amounts at the

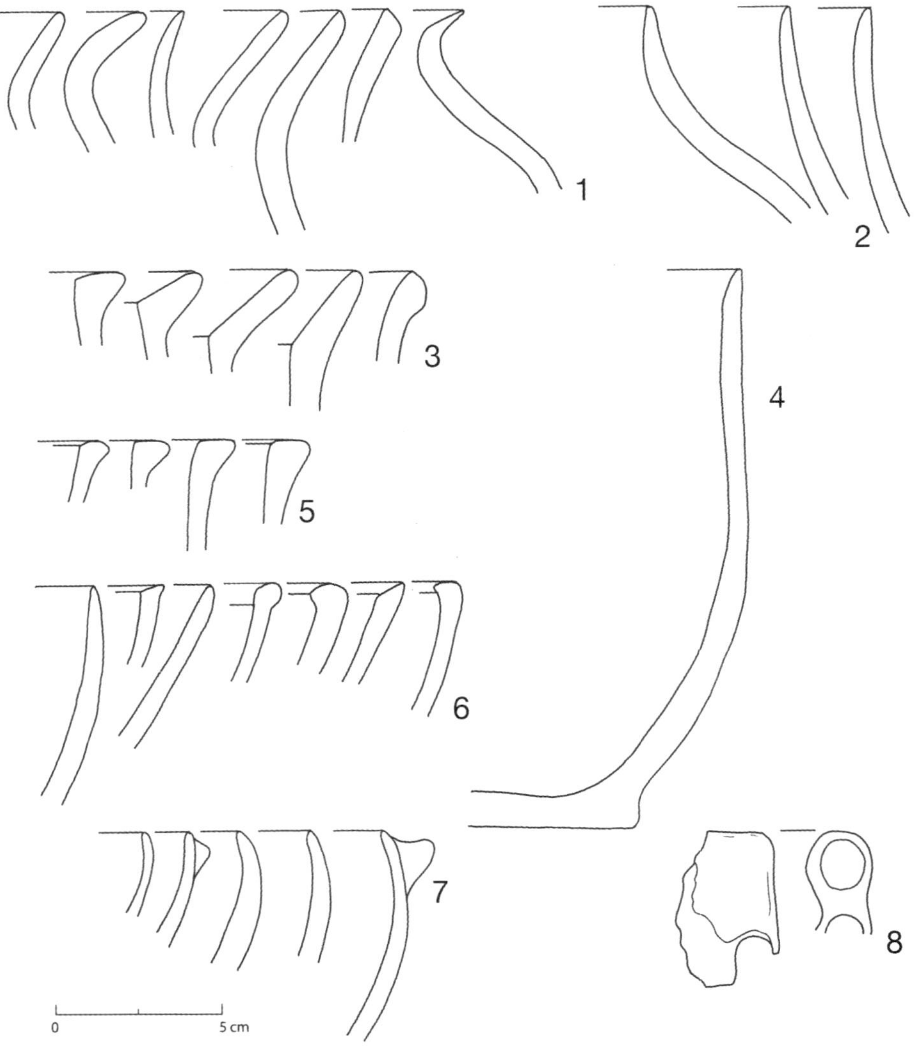

FIGURE 65. Bontour complex vessel forms and rim types from the Bontour, Quinam (Excavation 4) and St. Joseph 2 (Excavation 2) sites. (1) Bontour Form 1; (2) Bontour Form 2; (3) Bontour Form 3; (4) Bontour Form 4; (5) Bontour Form 5; (6) Bontour Form 6; (7) Bontour Form 7; (8) Bontour Form 8.

Bontour (1.1%) and St. Joseph 2 (2.0%) sites. Finally, at all sites a few potsherds seem to be tempered with pounded potsherds (grog). This is a tempering agent that dominates at the sites of the Marac assemblage in the central–southern part of Trinidad (see Boomert 1985, table 1). Note that some of the fine sand-tempered sherds encountered at Quinam, Excavation 4, may be ascribed to the Late Erin complex. Finally, both Bontour and Quinam, Excavation 4, yielded some potsherds tempered with river sand containing quartz, feldspar, and micaschist particles, obviously representing "trade" pieces from north Trinidad, perhaps St. Joseph 2.

SHAPE

Vessel Shapes. There are eight distinguishable vessel shapes; orifice sizes range from small to large (Figure 64). (The numbering of the vessel shapes identified here differs from that used previously for the combined Arauquinoid–Mayoid series in Trinidad [Boomert 1985]. Bontour Form 1 is identical to Arauquinoid–Mayoid Forms 3 and 5, Bontour Forms 2, 3, 4, and 5 to Arauquinoid–Mayoid Forms 4, 6, 7, and 8, respectively, Bontour Form 6 to Arauquinoid–Mayoid Forms 9, 10, and 11, and Bontour Forms 7 and 8 to Arauquinoid–Mayoid Forms 12 and 14, respectively.) Rims include direct, unmodified, and interiorly or exteriorly thickened forms next to typically inward-folded specimens. Three jar shapes, Forms 1, 2, and 3, have independent restricted orifices. Form 1 has inflected contours and a direct unmodified rim (Figure 65:1), Form 2 has simple contours and a direct, unmodified rim (Figure 65:2), and Form 3 has composite contours and a direct, interiorly or exteriorly thickened rim (Figure 65:3). Two jar shapes, Forms 4 and 5, have unrestricted orifices. Form 4 has simple contours and a direct unmodified rim (Figure 65:4). Form 5 is a "keeled" jar with composite contours and a direct unmodified rim, showing a concave profile above its corner point (Figure 65:5). Two bowl shapes, Forms 6 and 7, are known. Form 6 has an unrestricted orifice, simple contours, and a direct, flattened, interiorly thickened (beveled) or inward-folded rim (Figure 65:6). Form 7 has a restricted orifice, simple contours, and a direct unmodified rim (Figure 65:7). Finally, there is a bottle shape, Form 8, with a double independent restricted orifice, simple contours, and a direct unmodified rim (Figure 65:8). Several of the Bontour vessel shapes, notably Forms 3, 4, and 7, may have been used for the storage of dry goods or for food preparation without heating, while Forms 5 and 6 probably functioned as vessels for serving, drying, or displaying food. Jars most likely used for cooking include Forms 1 and 2, whereas the bottle shape of Form 8 clearly functioned as storage for liquids or as a traveling canteen. According to a representative sample of 828 rim sherds from Bontour complex sites all over Trinidad, including Bontour, St. Joseph 2, Excavation 2, and Quinam, Excavation 4, Form 6 predominates (53.5%), whereas Forms 7 (16.8%), 1 (14.8%), and 2 (10.6%) are less well represented. Forms 5 (2.4%), 4 (1.1%), 3 (0.7%), and 1 (0.1%) are rare (see Boomert 1985, table 2).

Base Forms. There are five base forms: flat with unmodified basal angle (A), flat with pedestaled basal angle (B), concave with unmodified basal angle (C), and annular and outflaring with unmodified basal angle (D). According to a representative sample of 497 base sherds from Bontour complex sites all over Trinidad, including Bontour, St. Joseph 2, Excavation 2, and Quinam, Excavation 4 (Boomert 1985, table 3), Base A is dominant (48.7%), followed by Base D (25.6%), and Base B (23.5%). Base C is rare (2.2%).

Handles. Handles encompass mainly vertical, D-shaped strap handles; similarly vertical, rod-shaped handles are rare.

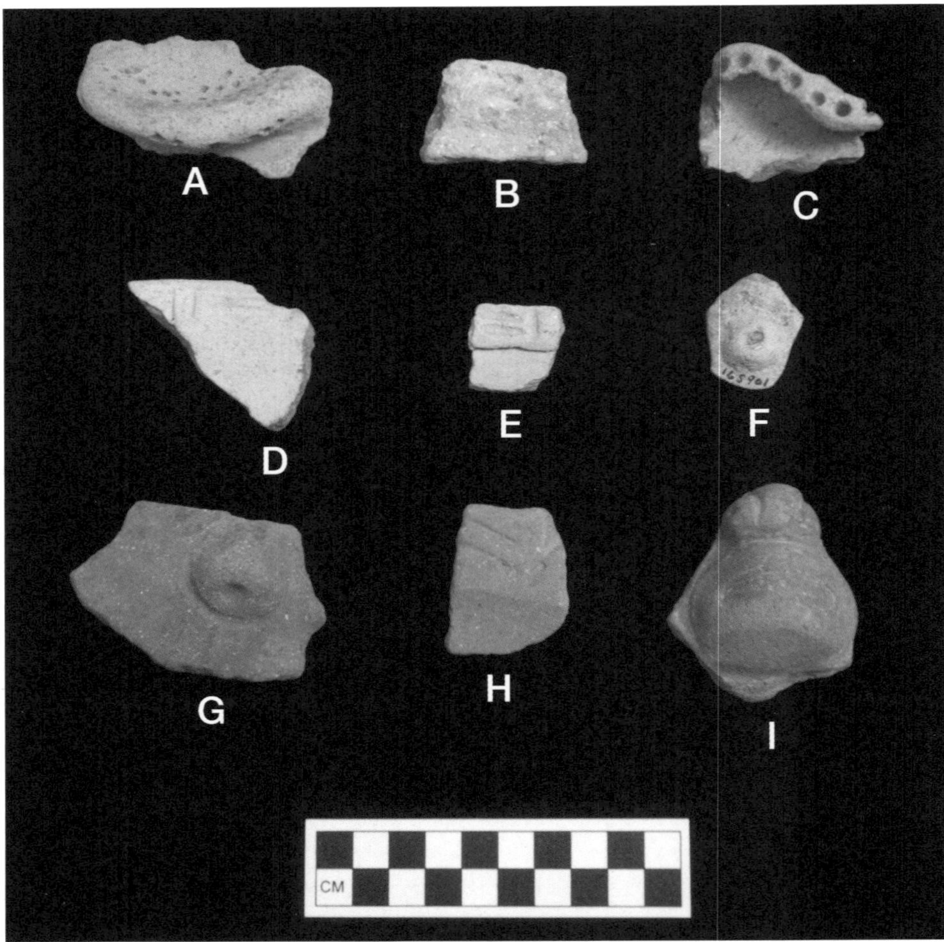

Figure 66. Incised and gouged, punctated, modeled-punctated, and modeled-incised pottery from the Bontour site (Carter's 1942 excavation and Excavation 1). **A–F,** Bontour complex. **G–I,** Erin complex. *Location finds:* A, B, D, G–I, Carter's excavation. C, Excavation 1, surface. E, Excavation 1, A1: 20–40 cm. F, Excavation 1, A2: 20–40 cm. *YPM Catalog Nos.:* (A) ANT 164999; (B) ANT 166153; (C) ANT 166194; (D) ANT 164987; (E) ANT 165862; (F) ANT 165901; (G) ANT 165041; (H) ANT 165040; (I) ANT 165042.

Decoration

Generally speaking, decoration is reduced to a minimum in the Bontour complex. At the sites of Bontour, St. Joseph 2, Excavation 2, and Quinam, Excavation 4, only 3.5%, 2.3%, and 2.8%, respectively, of the potsherds show some form of ornamentation. Five techniques of decoration can be distinguished: painting, incision (or gouging), punctation, simple modeling, and complex modeling.

Painting. Painting is rare, only 5.9% of a representative sample of 288 decorated potsherds from Bontour complex sites all over Trinidad, including Bontour, St. Joseph 2, Excavation 2, and Quinam, Excavation 4 (Boomert 1985, table 4). It is confined

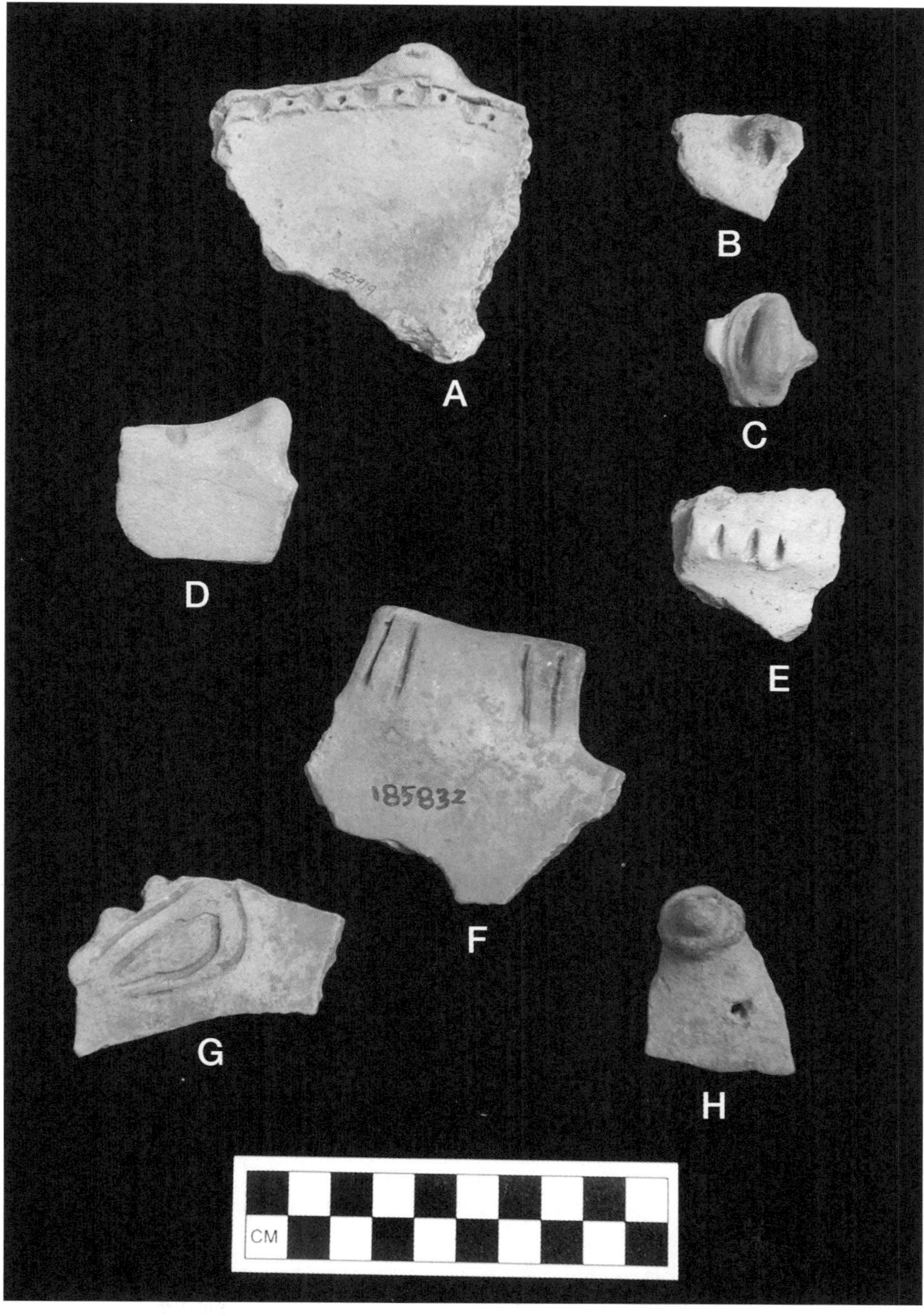

Figure 67. Punctated, incised, and modeled pottery of the Bontour complex from the St. Joseph 2 and Quinam (Excavation 4) sites. *Location finds:* A, St. Joseph 2, Excavation 2, A3: 30–40 cm. B, C, E, St. Joseph 2, Excavation 2, A1: 20–30 cm. D, St. Joseph 2, Excavation 1, A2: 10–20 cm. F–H, Quinam, Excavation 4, S2: 0–20 cm. *YPM Catalog Nos.:* (A) ANT 255919; (B, C, E) ANT 255810; (D) ANT 255660; (F) ANT 185832; (G) ANT 185833; (H) ANT 185918.

to the pre-fired application of monochrome red pigment all over the exterior of vessel surfaces.

Incision. In all, 26.7% of the decorated potsherds show some form of incision or gouging. The linear design elements are invariably wide and U-shaped in cross section, predominantly measuring about 1.5 to 2 mm across, although examples showing widths of up to 4 mm occur as well. Clearly, the narrowest lines were applied with a broad stylus with rounded end, but for the widest ones a small gouge made from a bird bone or hollow reed may have been used. Motifs include straight parallel lines (Figure 67A, F), single or multiple curved lines (Figure 67G), zigzags (Figure 67K), and friezes either of parallel rows of short incisions or gouges (Figures 62B, 66B) or of combinations of one or two parallel, horizontal lines (or gouges) alternating with one or two parallel, vertically placed lines. The latter motif is typically on the interiorly thickened or inward-folded rims of Form 6 bowls (Figures 66E, F and 67D).

Punctation. This is just as common as incision or gouging (25.4%). Motifs include straight or curving rows of punctations, occasionally associated with incised lines (Figures 66C and 67A). Fields of haphazardly dispersed punctations occur as well (Figure 66A). They may be combined with incised zigzags (Figure 62A).

Simple modeling. This is the most popular decorative technique (40.3%). Simple, round, or oval punctated rim and wall knobs and appendages represent the most frequently occurring designs (Figures 50C, 66A, F, and 67A–D), followed by small, triangular, or trapezoidal, rarely wavy, or "horned" rim lobes and raised rim lugs (Figure 67D, F). They sometimes have a central perforation, punctation, or incised circle. Punctated appliqué fillets, often triangular in cross section (Figures 66C, 67E), and dimpled rims are less common.

Complex modeling. Extremely rare (1.7%), complex modeled motifs include modeled-punctated geometric *adornos* (Figure 67H). Biomorphic head lugs have almost completely disappeared.

Other Kinds of Clay Artifacts

Four types of ceramic artifacts are known from the Bontour complex: griddles, a figurine foot, roller stamps, and an ocarina. The latter three artifact types are not known from Rouse's research at Bontour, St. Joseph 2, Excavation 2, and Quinam, Excavation 4 (see Boomert 1985). The griddles include platters provided with flat, unmodified rims, and specimens with thickened, upturned rims. They are often tempered with *cauixí* (Figure 67A).

Ceramic Age Nonpottery Artifacts

Most of the Cedros, Palo Seco, Erin, and Bontour objects made of materials other than pottery—notably stone, bone, and shell—functioned as tools related to the

Amerindian subsistence economy and ways of food processing. Because the adaptive strategies of the Ceramic complexes of Trinidad were essentially similar, based invariably on a combination of horticulture, hunting, fishing, and the collecting of plant and animal foods, the repertoire of nonpottery artifacts appears to be quite uniform. In addition, some of the objects represent bodily ornaments or may relate to the Amerindian religious convictions. Note that the discussion here is limited to the types of artifacts Rouse encountered at Cedros, Palo Seco, Erin, Quinam, Bontour, and St. Joseph 2, although these sites have yielded additional tool categories during other investigations.

BONE ARTIFACTS

Few tools and ornaments manufactured of animal bone were encountered during Rouse's excavations in Trinidad. These artifacts are associated with hunting and fishing, pottery manufacture, woodworking, bodily ornamentation, dress, and possibly shamanistic practices.

Bone tools connected with the hunting and fishing activities of the Ceramic Amerindians of Trinidad include a bipointed pencil fishhook (Figure 68B), found in Palo Seco context at Erin, Excavation 2 (S2: 100 to 120 cm). It resembles the Archaic specimens known from Banwarian Ortoiroid sites such as St. John (discussed above) and those used until recently by the Warao Indians of the Orinoco Delta, which were intended to be attached in the middle to a fishing line.

Detachable bone points for hunting or fishing have been found in Palo Seco–Erin context at Quinam, Excavation 1 (A3: 0 to 20 cm), and associated with Bontour ceramics at Bontour, both in Carter's excavation and Excavation 1 (A5: 0 to 20 cm), and at St. Joseph 2, Excavation 2 (A1: 20 to 30 cm).

A possible spearthrower spur (Figure 68G) is known from Palo Seco context at Palo Seco, Excavation 1 (A4: 40 to 60 cm). It may have been used both for hunting medium-sized mammals or, more likely, in warfare.

Several types of bone implements were probably used for craft activities. A small, rectangular chisel-like tool (Figure 68E) was found associated with Palo Seco pottery at Palo Seco, Excavation 1 (A4: 100 to 120 cm). It may have been an implement for incising pottery, but its use as a tattooing instrument is just as likely (see Boomert 2000:406).

Specific implements were clearly destined for the application of small adornments to loin cloths or feather headdresses and other piercing activities. Such tools include a small bone needle (Figure 68C) found in Palo Seco context at Palo Seco, Excavation 1 (A3: 80 to 100 cm), awls (Figure 68D), also found in Palo Seco context at Palo Seco, Excavation 1 (A3: 40 to 60 cm) and Excavation 2 (G3: 20 to 40 cm), and a sharpened deer horn (Figure 68I) recovered from Quinam, Excavation 1 (A3: 80 to 100 cm) and associated with Palo Seco and Erin pottery. Alternatively, the latter especially may have been a tool for prying large gastropods from their shells (see Boomert 2000:355, 407).

A deer bone tube decorated with several encircling incised lines (Figure 68H), from Palo Seco context at Quinam, Excavation 2 (G4: 40 to 60 cm), is a quite in-

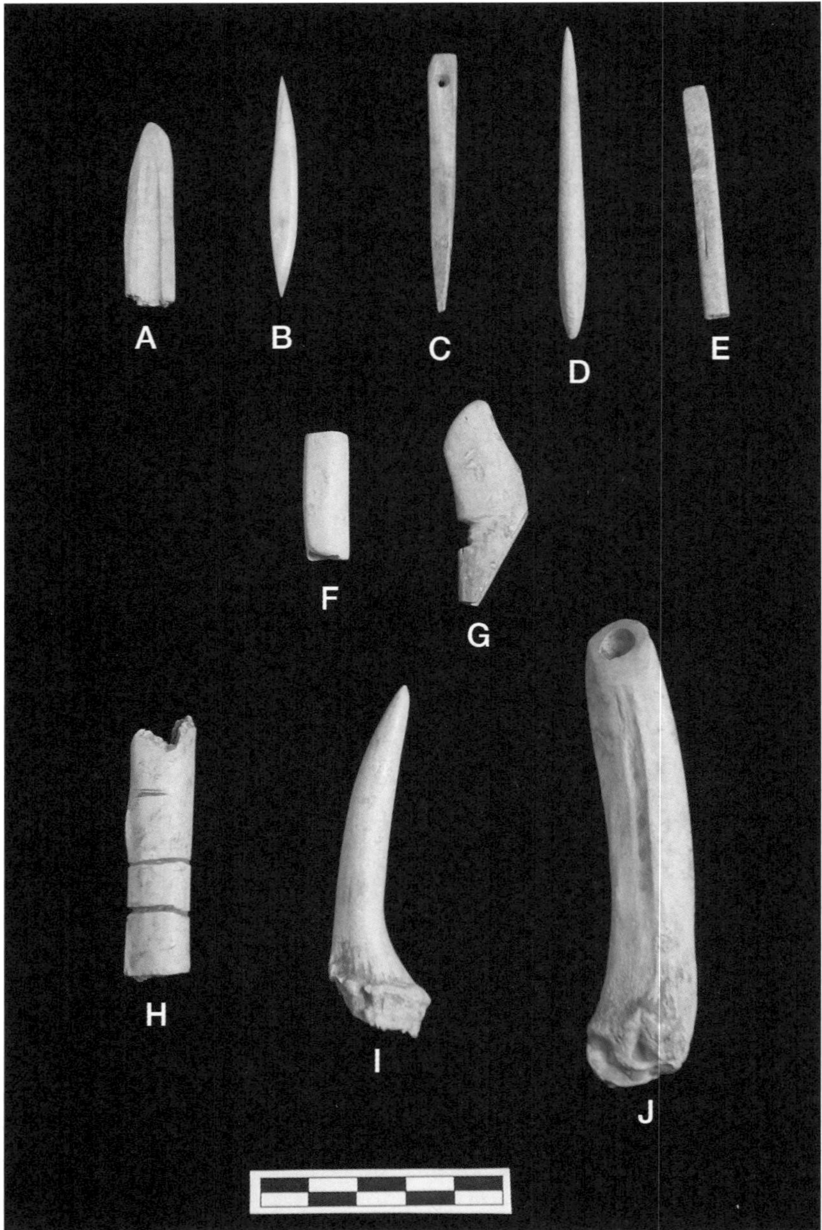

FIGURE 68. Bone artifacts from the Palo Seco, Quinam, and Erin sites. *Artifact types:* **A,** spatula fragment. **B,** bipointed fishhook. **C,** needle. **D,** awl. **E,** chisel-like tool possibly used for incising pottery or tattooing. **F,** tube. **G,** spearthrower spur (?). **H,** decorated tube. **I,** deerhorn with smooth point. **J,** cut deer bone. *Location finds:* A, Quinam, Excavation 1, A3: 60–80 cm. B, Erin, Excavation 2, S2: 100–120 cm. C, Palo Seco, Excavation 1, A3: 80–100 cm. D, Palo Seco, Excavation 1, A3: 40–60 cm. E, Palo Seco, Excavation 1, A4: 100–120 cm. F, Erin, Excavation 1, B2: 80–100 cm. G, Palo Seco, Excavation 1, A4: 40–60 cm. H, Quinam, Excavation 2, G4: 40–60 cm. I, Quinam, Excavation 1, A3: 80–100 cm. J, Quinam, Excavation 2, G3: 80–100 cm. *YPM Catalog Nos.:* (A) ANT 181273; (B) ANT 174118; (C) not numbered; (D) ANT 175417; (E) not numbered; (F) ANT 171190; (G) ANT 175508; (H) ANT 183452; (I) ANT 181567; (J) ANT 184302.

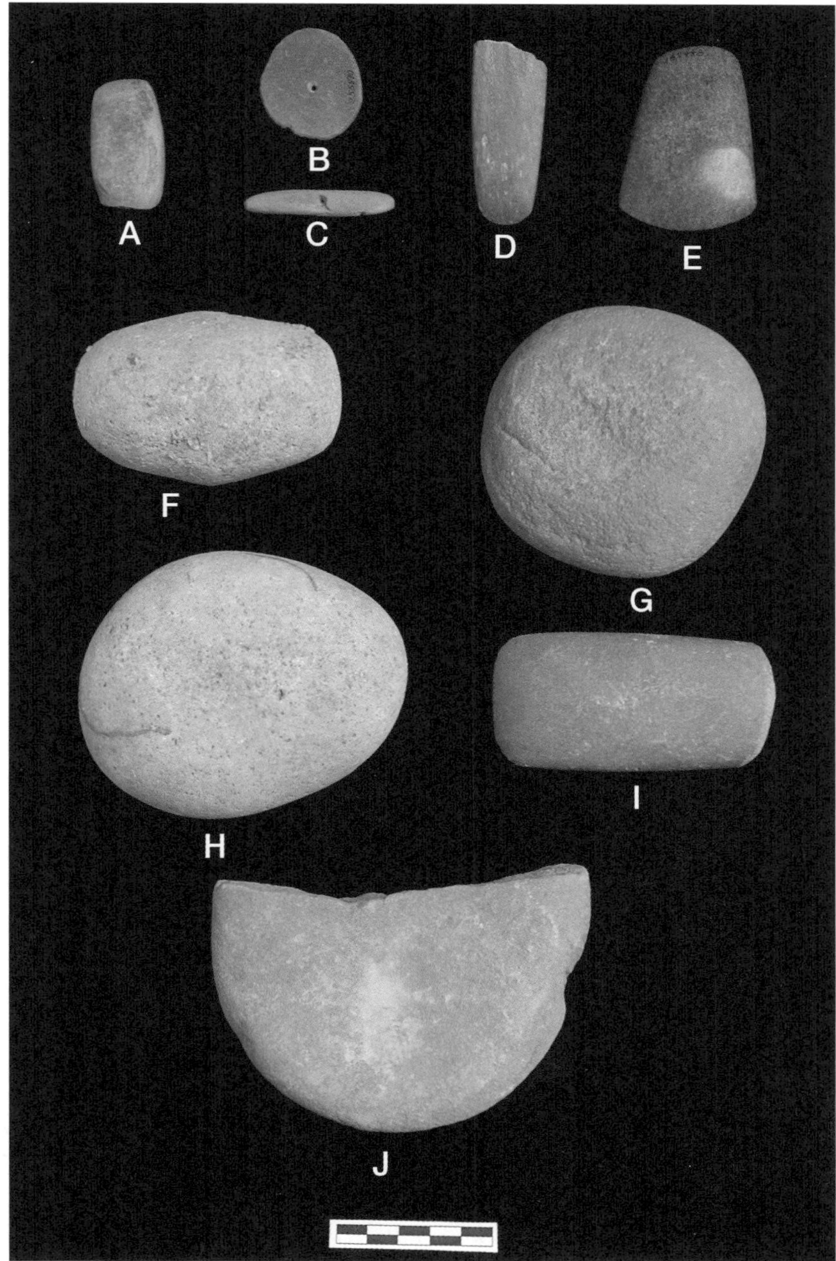

FIGURE 69. Stone artifacts from the Palo Seco, Quinam, and St. Joseph 2 sites. *Artifact types:* **A, D, E, I,** axe heads. **B, C,** pendants. **F,** pestle. **G, H,** anvils. **J,** grinding stone. *Location finds:* A, Palo Seco, Excavation 1, A4: 40–60 cm. B, St. Joseph 2, Excavation 2, A2: 40–50 cm. C, St. Joseph 2, Excavation 2, A1: 10–20 cm. D, Quinam, Excavation 1, A1: 20–40 cm. E, Quinam, Excavation 2, G4: 40–60 cm. F, Quinam, Excavation 2, G6: 60–80 cm. G, Quinam, Excavation 1, A2: 20–40 cm. H, Quinam, Excavation 2, G5: 80–100 cm. I, Quinam, Excavation 1, A2: 60–80 cm. J, Quinam, Excavation 2, A2: 60–80 cm. *YPM Catalog Nos.:* (A) ANT 175509; (B) ANT 255870; (C) ANT 255806; (D) ANT 180307; (E) ANT 183448; (F) ANT 184214; (G) ANT 180394; (H) ANT 184623; (I) ANT 181426; (J) 183447.

triguing object. Comparable, though undecorated, tube-shaped bone artifacts were encountered at Cedros, Excavation 1 (A5: 20 to 40 cm), associated with transitional Cedros–Palo Seco pottery (Faber- Morse 2007), and at Erin, Excavation 1 (B2: 80 to 100 cm), in correlation with predominantly Erin ceramics. These bone tubes seem to be too small and thin to have been mouth pieces for cigars used by shamans during healing practices, such as are known from Amazonia and Oronoquia. However, they may have served various comparable purposes, such as tubes for pouring tobacco or pepper juice into the nostrils to induce an ecstatic-visionary trance, to blow tobacco smoke over a patient, as enema tubes, or to suck "spirit stones" out of a patient (see Boomert 2000:480).

Implements of which the function is unclear include a spatula-like tool (Figure 68A) found in Palo Seco–Erin context at Quinam, Excavation 1 (A3: 60 to 80 cm), and a cut deer bone (Figure 68J) from Quinam, Excavation 2 (G3: 80 to 100 cm) found associated with Palo Seco pottery. A similar cut piece of deer bone is known from transitional Cedros–Palo Seco context at Cedros, Excavation 1 (A1: 0 to 20 cm).

Bone ornaments, finally, include beads and pendants. A bone bead (Figure 68F) was found at Erin, Excavation 1 (B2: 80 to 100 cm), associated predominantly with Erin pottery. A clearly male-associated category of ornaments is that of perforated or (yet) undrilled peccary tusks. Such an unperforated peccary tusk is known from Bontour context at Bontour, Excavation 1 (A1: 0 to 20 cm), and two specimens (one drilled) were recovered from Quinam, Excavation 2 (G3: 40 to 60 cm), associated with Palo Seco pottery.

STONE ARTIFACTS

The stone tools encountered during Rouse's excavations in Trinidad encompass artifacts associated with horticulture, hunting and fishing, food processing, pottery manufacture, and other craft activities, as well as bodily ornamentation.

All sites have yielded polished stone axe heads (Figure 69A, D, E, I). These are single-edged, generally moderate in size (about 10 to 12 cm in length), and show a trapezoidal ("petaloid") shape. Attached to wooden hafts, this type of celt was used for felling trees to obtain posts for house construction and for clearing swidden plots for horticulture, whereas specimens with somewhat asymmetrical cross sections may have been used for canoe manufacture. Chisel-like forms with almost parallel sides, driven by hammerstones, could have served primarily for specialized woodworking. Trees were cut by alternately charring the trunk with fire and cutting away the burned wood. Consequently, the stone axes were used as bruising rather than as cutting instruments (see Boomert 2000:297, 315–317, 336). Polished stone axe heads are known from transitional Cedros–Palo Seco context at Cedros, Excavation 1 (A1: 0 to 20 cm), and associated with Palo Seco ceramics at Palo Seco, Excavation 1 (A4: 40 to 60 cm) and Excavation 2 (G1: 0 to 20 cm and G3: 0 to 20 cm), Quinam, Excavation 2 (G4: 40 to 60 cm), and Erin, Excavation 1 (C4: 120 to 140 cm). They have been found associated with Palo Seco and Erin pottery at Quinam, Excavation 1 (A2: 60 to 80 cm), and with predominantly Erin ceramics at Quinam, Excavation 1 (A1: 20 to 40 cm), and Erin, Excavation 1 (B1: 80 to 100 cm). Finally, stone celts are known from Bontour context at Bontour, Carter's excavation and Excavation 1 (A1: 0 to 20 cm), and St. Joseph 2, Excavation 2 (A1: 10 to 20 cm). Note that most of the stone axe heads are made from metamorphic rock, most likely originating in the Venezuelan Coastal Cordillera, Tobago, and the Lesser Antilles (see Boomert 2000:436–437).

A few stone artifacts are connected with the hunting and fishing activities of the Trinidadian Amerindians of Ceramic times. These include a sandstone arrow-shaft polisher found at Quinam, Excavation 1 (A3: 20 to 40 cm), associated with predominantly Erin ceramics and two crude notched stones that can be interpreted as net weights, thus suggesting the use of seines for fishing, encountered at Quinam, Excavation 2 (G4: 40 to 60 cm and G6: 60 to 80 cm), associated with Palo Seco pottery. Stone objects functioning as food processing tools include pestles, anvils, grinding stones, and coral rasps or graters. A pestle (Figure 69F), found at Quinam, Excavation 2 (G6: 60 to 80 cm), may have been used to pound root crops other than cassave (for example, to mash sweet potatoes and moist or leafy vegetable foods).

Anvils (Figure 69G, H) or "pitted stones" were used with small hammerstones to crack palm nuts. Specimens are known from Palo Seco context at Quinam, Excavation 2 (G5: 80 to 100 cm), and Erin, Excavation 1 (B1: 140 to 160 cm), and associated with predominantly Erin pottery from Quinam, Excavation 1 (A2: 20 to 40 cm) and Erin, Excavation 1 (B1: 80 to 100 cm).

Grinding stones (Figure 69J) have been found associated with Cedros ceramics at Cedros, Excavation 1 (A1: 40 to 60 cm), in transitional Cedros–Palo Seco context at Cedros, Excavation 1 (A1: 20 to 40 cm), and Palo Seco, Excavation 2 (G2: 40 to 60 cm), and, finally, associated with Palo Seco pottery at Palo Seco, Excavation 2 (G3: 0 to 20 cm) (Faber-Morse 2009), and Quinam, Excavation 2 (G4: 40 to 60 cm and A2: 60 to 80 cm). They may have been used to grind wild seeds, but could as well have been used partially for grinding and polishing stone, shell, bone, and wooden implements and ornaments. Pieces of fossil coral were probably used for similar purposes next to the grating of cassava and other root crops. Such coral rasps have been found in Palo Seco context at Quinam, Excavation 1 (A4: 100 to 120 cm). They must have formed items of exchange from north Trinidad or Tobago, because neither living nor fossil coral is found along Trinidad's south coast.

Some stone tools were used for craft activities. Polishing stones and hematite rubbing stones were part of the pottery manufacturing process. Polishing stones were used to burnish or smoothen a pottery vessel when the clay was still leather-dry. They are known to be associated with Cedros pottery at Cedros, Excavation 1 (A4: 40 to 60 cm), in transitional Cedros–Palo Seco context at Cedros, Excavation 1 (A2: 0 to 20 cm and A2: 20 to 40 cm) (Faber-Morse 2007), associated with Palo Seco ceramics at Erin, Excavation 1 (C4: 160 to 180 cm), and Quinam, Excavation 1 (A3: 100 to 120 cm) and Excavation 2 (G3: 40 to 60 cm and G4: 60 to 80 cm), and, finally, with predominantly Erin pottery at Erin, Excavation 1 (A2: 80 to 100 cm).

Rubbing stones of red ochre (hematite) were clearly the source of much of the red pigment on the Saladoid pottery of the Caribbean that was possibly mixed with vegetable gum (see Boomert 2000:142). Hematite has been found in appreciable quantities at most Ceramic sites in Trinidad, so its presence in Rouse's excavations will not be detailed here. Artifacts possibly used for craft activities such as woodwork, basketry, and plaitwork include small, irregular stone flake and core tools of chert, flint, quartz, quartz crystal, quartzite, and other local rock materials. Many types of these unsophisticated small implements are known from the Ceramic sites of Trinidad and their presence in Rouse's excavations also will not be detailed here. Manufactured by simple, freehand percussion flaking or the bipolar technique, these unmodified expedient tools often still show remnants of the cortex. They include flake scrapers, cutters, and knives, some of which may have been used to process plant fibers for basketry, along with the cutting, sawing, scraping, and shaft shaving of wood. In general, only some 10% of these flakes and chips show use wear. Small, round to oval or slightly tapered hammerstones were used to produce these flake and core implements. Examples are known from transitional Cedros–Palo Seco context at Cedros, Excavation 1 (A3: 20 to 40 cm), associated with Palo Seco pottery at Palo Seco, Excavation 2 (G1: 0 to 20 cm and G3: 0 to 20 cm), and Erin, Excavation 1 (B1: 120 to 140 cm), and with Erin ceramics at Quinam, Excavation 1 (A1: 20 to 40 cm).

Stone ornaments, finally, include beads and pendants. Stone beads have been found in Palo Seco context at Quinam, Excavation 3 (M1: 20 to 40 cm), and associated with Bontour pottery at Bontour, Excavation 1 (A2: 0 to 20 cm). The Bontour specimen is button-shaped and made of diorite, indicating that it was an exchange item possibly originating from Tobago. Note that an unworked piece of amethyst, a similarly exotic rock type, was encountered at Cedros during the 1969 excavations at the site by Rouse, Cruxent, and Olsen. It was found associated with Cedros pottery in level 3 (50 to 75 cm) of their Pit 1 (see Boomert 2000:410; Harris, pers. comm. to Boomert 1983). In addition, a button-shaped bead of amethyst was recovered in Cedros context from Pit B1, level 10 (45 to 50 cm), during the 2005 excavations at the Cedros site. Three stone pendants are known from St Joseph 2, Excavation 2 (A1: 10 to 20 cm, A2: 30 to 40 cm, and A2: 40 to 50 cm), all associated with Bontour ceramics (Figure 69B, C). Two specimens, of which only one is perforated, are circular, whereas a third pendant is elliptical with an unfinished transfixion. All three pendants are made of thin slabs of micaschist, a shiny rock type typical of Trinidad's Northern Range.

SHELL ARTIFACTS

The only tool made of shell was found in Bontour context at Bontour, Excavation 1 (A3: 0 to 20). It is a small fragment of an axe, possibly intended for making canoes, manufactured from an unidentified gastropod. All other shell artifacts are finished or unfinished pendants made from the valves of freshwater mussels (Unionidae) from which the external layer has been removed to expose the brilliant nacre underneath. Unworked valves of these naiad shells have been found associated with Cedros pottery at Palo Seco, Excavation 2 (G5: 120 to 140 cm) and in transitional Cedros–Palo Seco context at Palo Seco, Excavation 2 (G5: 40 to 60 cm). They are

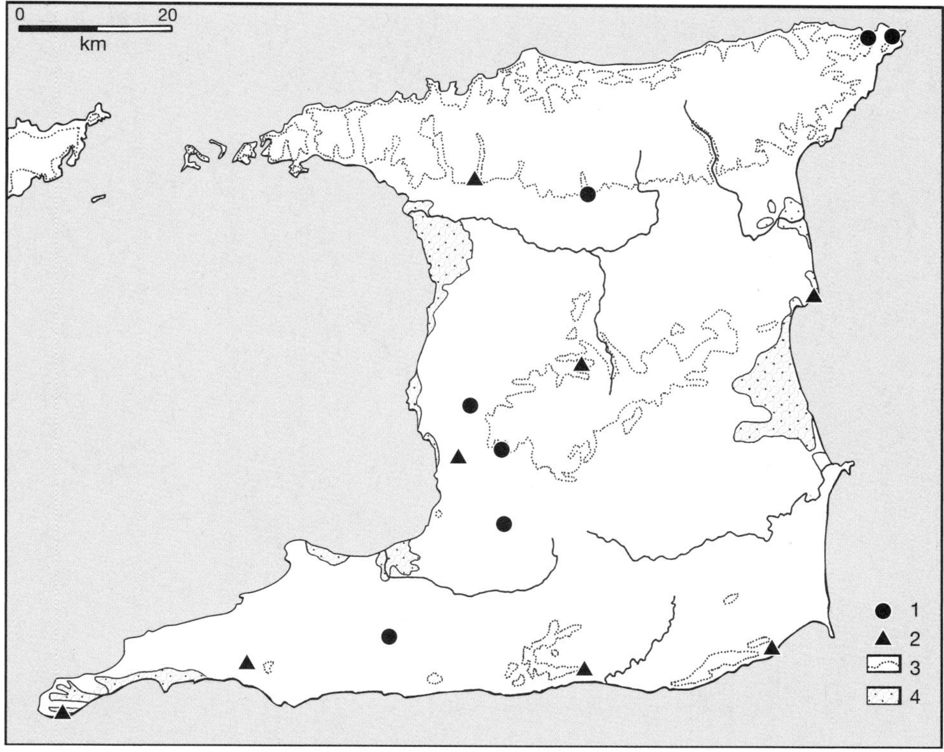

Figure 70. The location of Mayo complex sites on Trinidad. *Legend:* (1) Spanish-Amerindian mission sites; (2) other sites; (3) 100 m contour; (4) swamps and marshes.

also known from Palo Seco context at Palo Seco, Excavation 1 (A1: 40 to 60 cm, A2: 40 to 60 cm, A3: 120 to 140 cm, and A6: 20 to 40 cm), Palo Seco, Excavation 2 (G3: 20 to 40 cm), and Erin, Excavation 2 (S2: 100 to 120 cm and S2: 120 to 140 cm). Another example was encountered at Quinam, Excavation 1, associated with Palo Seco and Erin ceramics, but its stratigraphic origins were not recorded.Note that a semi-circular, perforated specimen of such a naiad shell pendant was excavated by Bullbrook at the Erin site (see Boomert 2000, fig. 65:3) and an unworked example was found in transitional Cedros–Palo Seco context during the 2005 excavations in Pit B2 (levels 3 and 4, 5 to 15 cm) at the Cedros site.

Ceramic–Historic Age Complex

The final Amerindian tradition of Trinidad, the Mayoid series, is characteristic of the Amerindian–European contact period. It may have emerged shortly before the time of Columbus's discovery of the island in 1498, the result of immigration by Amerindians from the coastal zone of the Guianas. Although Mayoid ceramics were first recovered by Rouse and Goggin at the Mayo site in 1953, they were erroneously interpreted as part of the Bontour complex (see Rouse 1953). It was not until the 1980s that the Mayoid series was first defined (Boomert 1985). Mayoid

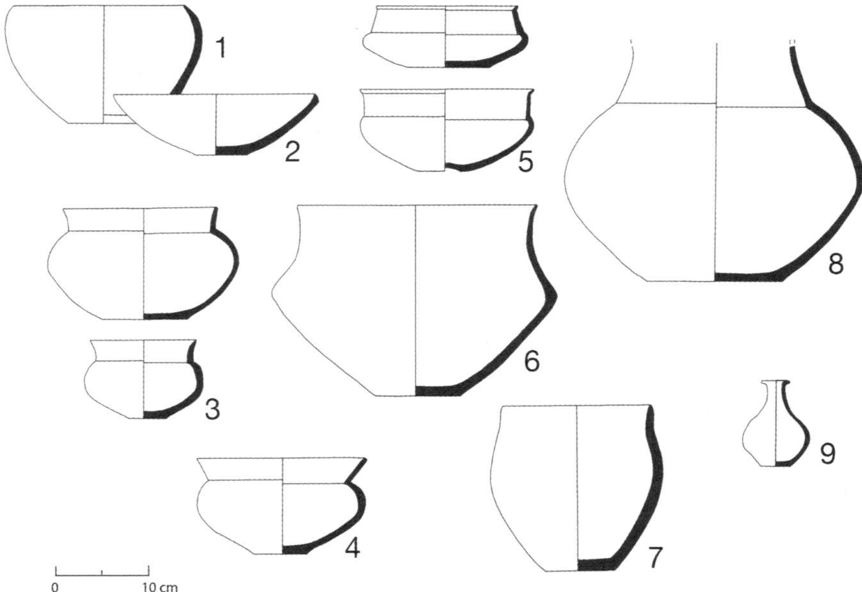

FIGURE 71. Reconstructed Mayo complex vessel forms from Trinidad (all except Form 8 to the same scale).

ceramics typify the sites of the Capuchin missions established by the Spanish from 1687 onward and may have been made until as late as the mid-eighteenth century.

Mayo (Mayoid)

Pottery of the Mayo complex has been found at fifteen sites in Trinidad (Figure 70), including seven seventeenth- to eighteenth-century Spanish-Amerindian mission sites (Boomert 1985, 1986). In addition, it characterizes the Amerindian pottery encountered with Spanish *majolica* and other wheel-made historic period ceramics at the town of San José de Oruña, Trinidad's Spanish capital until 1784. This suggests that the Spanish used the Mayo pottery as kitchenware. One of the Capuchin mission sites has yielded human burials (Boomert 1985). The remaining eight sites include former Amerindian settlements and camp or bivouac sites, at least some of which also date from the Historic age. Note that the previously hypothesized distinction between two Mayoid ceramic complexes, Mayo and Guayaguayare, is discontinued here. Sites are evenly distributed throughout the entire island. Without doubt the Mayo earthenware was made by the Indians of the Nepoio and Arawak ethnic groups in Trinidad and perhaps by other Amerindians as well. In 1953, Rouse and Goggin first encountered Mayo pottery at the Mayo site, which they were able to identify as the site of the mission of Nuestra Señora de Montserrate (dated from 1700–1705 to 1789). Mayo pottery associated with European wheel-made wares typify Goggin's three trenches, Excavations 1, 2, and 3, at the Mayo site. It was also found together with Spanish *majolica* throughout Excavation 1 at St. Joseph 2 and in the upper portion, levels 1 and 2 (0 to 40 cm), of Excavation 2 at this site.

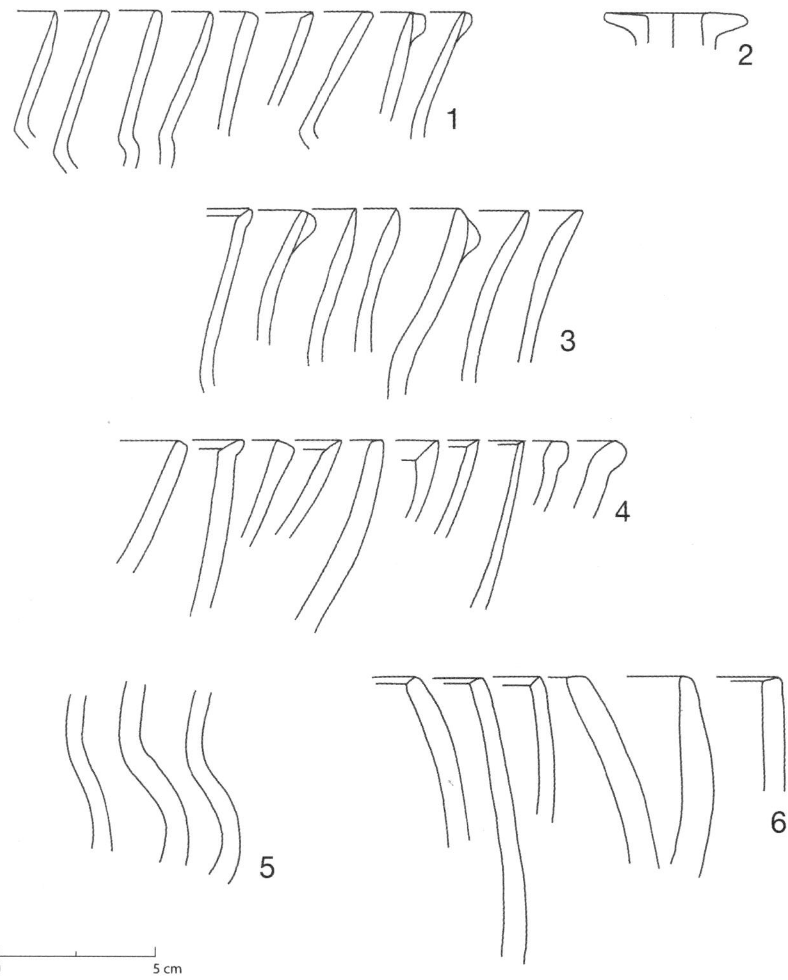

FIGURE 72. Mayo complex vessel forms and rim types from the Mayo site. *Legend:* (1) Mayo Form 4; (2) Mayo Form 9; (3) Mayo Form 3; (4) Mayo Form 2; (5) Mayo Form 7; (6) Mayo Form 1.

MATERIAL

Vessels were made by coiling and were fired in an open fire. Vessel walls are relatively thin; thickness ranges from 4 to 10 mm, averaging 4 to 6 mm. Surfaces are generally well smoothed. Fire clouds are common. The pottery is medium-hard, about 2.5 to 3 on the Mohs scale. Oxidization is incomplete. Sherds are predominantly yellowish gray to tan and dark gray in cross section. Mayo pottery is invariably tempered with *caraipé*, the ash of the siliceous bark of small trees belonging to the *Licania* genus (Chrysobalanaceae). This is known among the Amerindians of the Amazon Valley and the Guianas as *couepia* (*kwepi, kwep*) or *kauta*. At present it is the only tempering material the Indians of the littoral part of the Guianas use for their pottery. The bark is burned, removing most organic components, and afterward pounded. Temper consists of a mixture of white to gray siliceous particles, columnar and cellular in structure, and grains of carbonized organic material. The *Licania* tree was once indigenous in the savannahs of Trinidad's Northern Basin. The temper particles are occasionally quite coarse, up to 6 mm in length, and well visible on the exterior vessel walls (Figure 73D, E). Finally, potsherds tempered with river sand containing quartz, feldspar, and micaschist particles, obviously representing "trade" pieces from north Trinidad, possibly Arima, were recovered from Excavation 1 (A1: 0 to 20 cm) and Excavation 3 (A5: 20 to 40 cm) at the Mayo site.

SHAPE

Vessel Shapes. There are nine distinguishable vessel shapes with orifice sizes ranging from small to large (Figure 71). By far, most vessel rims are direct, typically showing flattened or beveled lips, or both. Two bowl shapes, Forms 1 and 2, are known, showing an unrestricted and restricted orifice, respectively. Both vessels have simple contours (Figure 72:4, 6). The most characteristic Mayo vessel comprises two jar shapes, Forms 3 and 4, closely resembling the Arawak (Lokono) cooking vessel or "buck-pot" of the Guianas. Both variants show an independent restricted orifice, composite contours, and a thin-walled, sharply everted neck that is either slightly outcurving (Form 3; Figure 72:3) or straight (Form 4; Figure 72:1). Form 5 is a small jar with dependent restricted orifice and composite contours. This vessel shape is almost identical to the necked jar characteristic of the late-prehistoric to historic period Koriabo and Cayo pottery complexes of the Guianas and Windward Islands, respectively (Boomert 1986). Form 6 is a sizeable jar with dependent restricted orifice and composite contours. Two other jar shapes, Forms 7 and 8, show independent restricted orifices and inflected and composite contours, respectively. The profile of Form 8 is not entirely known; it is a huge vessel with a diameter of 39.5 cm at a height of 6.5 cm. Finally, Form 9 is a bottle-like shape with an independent restricted orifice, an exteriorly thickened rim, and composite contours (Figure 72:2). According to a representative sample of 392 rim sherds from Mayo complex sites (Boomert 1985, table 2), Form 3 is most common (33.9%), whereas Form 4 (28.1%) and Form 2 (25.2%) range second and third. Form 1 (12.5%) is less well represented and Forms 5, 6, 7, 8, and 9 are rare. Several of the Mayo vessel shapes, notably the bowls and jar of Forms 1, 2, and 5, may have functioned as vessels for serving, drying, or displaying food, while the jar of Form 7 may have been used for the storage of dry goods or for food preparation without heating. Forms 3 and 4 were obviously cooking jars, whereas the bottle shape of Form 9 clearly functioned as storage for liquids or as a traveling canteen. The sizeable jars of Forms 6 and 8 most likely were containers for the storage and fermentation of cassava beer.

Base Forms. There are five distinguishable base forms (Boomert 1985, fig. 7): flat with unmodified basal angle (A), flat with pedestalled basal angle (B), concave with unmodified basal angle (C), and annular and outflaring with unmodified basal angle (D). According to a representative sample of 322 base sherds from Mayo complex sites (Boomert 1985, table 3), Base A is dominant (73.3%), followed by Base B (25.6%). Bases D and C are rare.

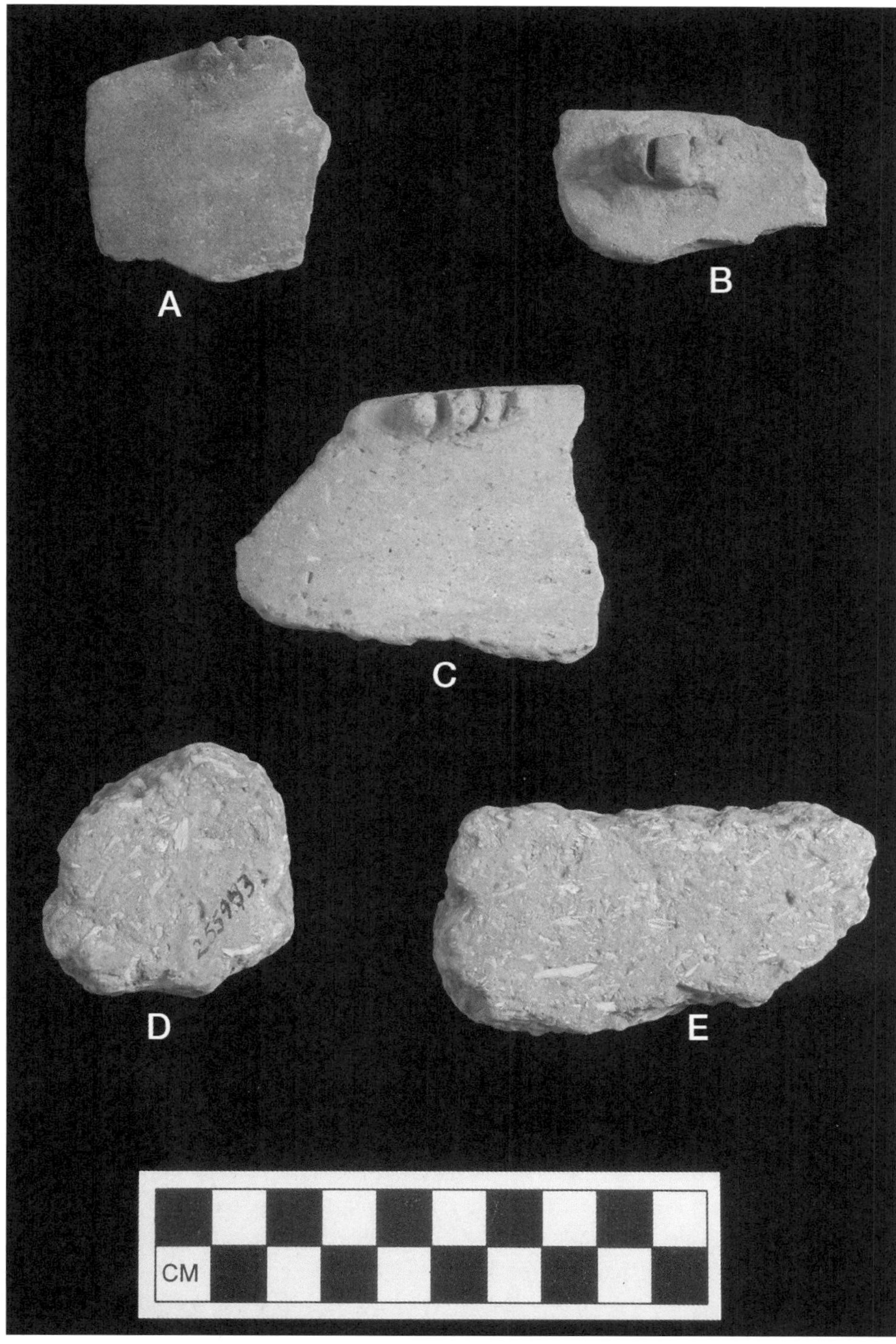

FIGURE 73. Decorated and undecorated pottery of the Mayo complex from the Mayo site. *Location finds:* A, C–E, Surface. B, Excavation 2, A1: 0–20 cm. *YPM Catalog Nos.:* (A) ANT 255936; (B) ANT 255995; (C) ANT 255942; (D) ANT 255963; (E) ANT 255943.

Handles. Handles are exceptional, encompassing only vertical, D-shaped rod handles.

Decoration

Pottery decoration is even further reduced now: at the Mayo site only 31 (1.3%) of the 2,342 potsherds recovered during the excavations from 1953 to 1978 show some form of ornamentation. Four techniques of decoration can be distinguished: painting, incision, punctation, and simple modeling.

Painting. Painted decoration is extremely rare, including the pre-fired application of monochrome red pigment all over the interior or exterior, or both, of vessel surfaces and the occurrence of narrow black painted zones along the vessel rims, especially in Forms 3 and 4. In addition, the interior and exterior vessel walls occasionally show traces of black smudging (Boomert 1986, fig. 21:5–6, 8).

Incision. Incised decoration is exceptional and confined to a single horizontally incised line below a vessel rim (Boomert 1985, fig. 14, B:3).

Punctation. Punctation is the second most common decorative technique (34.6%), restricted to punctated or nicked impressions of simply modeled knobs and pellets (Figure 73A–C).

Simple modeling. The most popular decorative technique (59.3%) is simple modeling. Small round or oval, punctated, nicked or undecorated rim and wall knobs are most frequent. Diminutive rim lobes are rare (Boomert 1985, fig. 14C:3–7).

Other Kinds of Clay Artifacts

Griddles form the only type of ceramic artifacts known from the Mayo site. They include platters provided with flat, unmodified rims and specimens with thickened, raised rims (Boomert 1985, fig. 16, A:5, B:9). The underside of some specimens show the impressions of the large leaves on which the griddle was formed before firing.

Nonpottery Artifacts

The Mayo site yielded a series of stone artifacts, including a rectangular stone axe head (Excavation 1, A1: 0 to 20 cm), a grinding stone (Excavation 1, A1: 0 to 20 cm), a hammerstone (Excavation 1, A1: 0 to 20 cm), a polishing stone for pottery (Excavation 1, A1: 20 to 40 cm), three hematite rubbing stones (Excavation 1, A1: 0 to 20 cm), and several small chert chips and flakes (Excavation 1, A1: 0 to 20 cm and A2: 0 to 20 cm). The presence of these stone implements at the site of a Spanish–Amerindian mission points in some degree to the continuation of a "prehistoric" pattern of subsistence and food procession. This is confirmed by the archaeozoological remains recovered from the site (see Chapter 4).

CHAPTER FOUR

ARCHAEOZOOLOGICAL FINDS

The animal and fish bones recovered by Rouse and Goggin allow a general insight into the adaptive strategies of the prehistoric through protohistoric Amerindian populations of Trinidad. Although because of rain during the excavation work no fine-mesh screens could be used, the archaeozoological finds encountered suggest characteristic trends in hunting, fishing, and food collecting from as early as Archaic times to as late as the eighteenth century. The outline presented here is primarily based on the analyses of Rouse's 1946 and 1953 finds by Wing (1962:33–52, 1977) and Wing and Reitz (1982), which, partially in the light of newly recovered evidence, were revised by Harris (1973, 1976) and Boomert (2000:57–58, 61–64, 339–344).

Archaic Age Finds

The fish and other animal remains recovered from the St. John shell midden accurately reflect the Archaic subsistence adaptations of the site's former occupants. They point to an Amerindian "broad-spectrum" diet accomplished through specialized modes of hunting, fishing, and food collecting, the latter primarily the gathering of shellfish, crabs, and edible wild plants, nuts, and berries. Only the archaeozoological materials encountered by Rouse at St. John (especially the mammalian remains) have been sufficiently analyzed (Wing 1962:33–38, 1977). However, misinterpretation of the site's stratification—which, as noted above, showed late-prehistoric Bontour ceramics in its top levels in apparent mechanical admixture with the preceramic deposit—led Wing to the erroneous conclusion that the St. John shell midden can be divided into a lowermost portion dating from Archaic times and an upper zone accumulated in the latter portion of the Ceramic age, approximately corresponding to Stratum A and Stratum B, respectively (also Newson 1976:37, 48, 52–53). Neither the artifact assemblage nor the sequence of vertebrate food remains at St. John supports such a conclusion and, consequently, Wing's analysis should be modified accordingly. The results of Harris's subsequent 1972 investigation at St. John can be used to broaden our understanding of the Archaic food-getting strategies at the site. The same applies to his analysis of the invertebrate fauna encountered at Ortoire (Harris 1976). Unfortunately, the animal bone remains of Ortoire have been insufficiently analyzed. The importance of the collecting of wild plant foods at both St. John and Ortoire can be inferred mainly from the presence of specialized ground stone artifacts, including grinding stones and pitted anvils. Unidentified charred nuts are reported only from Rouse's excavations at Ortoire (Boomert 2000:63-64, 85).

The vertebrate remains from St. John reflect a significant difference in the intensity of habitat exploitation between Stratum A, the bottom zone of refuse deposition at the site, and Stratum B, the middle layer of the site's stratification. (The upper zone, Stratum C, was not recovered in Rouse's Excavation A.) There was apparently a major change in food-getting strategies used for the two most extensively exploited ecosystems, the terrestrial (forest) habitat and the marine, inshore–estuarine environment. While in all 58.4% of the vertebrate remains from Stratum A derive from species at home in the terrestrial ecosystem, this figure drops to 33.4% during the deposition of Stratum B. In contrast, the importance of species deriving from the marine inshore–estuarine habitat increases simultaneously from 35.6% to 54.4%. Vertebrate remains from the marine banks and reefs as well as the offshore pelagic and freshwater ecosystems remain negligible, although it appears to be significant that the latter two habitats were completely unexploited during the occupation represented by Stratum A (Boomert 2000, table 3). An increase in the use of the aquatic biotic zone (fishing) as opposed to that of the terrestrial habitat (hunting) seems to be indicated. This pattern duplicates that at the Banwari Trace site, investigated by Harris (1973, 1976), which could be correlated with the ecological alteration of the Oropuche Lagoon from a freshwater or slightly brackish lagoon to a marine mangrove swamp resulting from the full submergence that created the Gulf of Paria by around 6200 BP (about 5100 cal BC), when the sea reached its present level. Apparently, during the occupation of St. John represented by Stratum A (Early Banwari Trace) the shore of the Gulf of Paria was much farther from the site than it is today and has been from Stratum B (Late Banwari Trace) onward (see also Wing 1977).

At St. John the most important animals hunted by the local Amerindians included collared peccaries (29.5%), nine-banded armadillos (14.0%), pacas (13.5%), and red brocket (12.4%), all forest-loving mammal species. Black-eared opossums, red howler monkeys, tree rats, two species of spiny rats, prehensile-tailed porcupines, agoutis, crab-eating raccoons, otters, and ocelots are less abundantly represented. No particular change in species preference is noted between Stratum A and Stratum B. The hunting of birds (exclusively pygmy owls and muscovy ducks), amphibians (toads), and reptiles (crocodiles, tortoises, musk turtles, snakes, and tegus) was insignificant. Almost half of the fishes caught in the preferred ecosystem, the marine inshore–estuarine habitat of the Gulf of Paria, belong to the sea catfishes family. Other fishes at home in this particular ecosystem represented in the deposit include spotted eagle rays, ladyfishes, toadfishes, snooks, jacks, blue runners, crevalle jacks, tripletails, tarpons, sea trouts, mullets, and flounders. The recovery of sparse remains of groupers, snappers, and grunts points to limited exploitation of the marine banks and reefs, that of mackerels and tunas to exceptional fishing in the offshore pelagic waters (Wing and Reitz 1982; Boomert 2000, table 2). The mudflats of the Gulf of Paria and the mangroves of the Oropuche Lagoon were the primary shellfish collecting areas of St. John. Tiger lucinas and nerites were gathered in the former area, Caribbean oysters and West Indian Crown conchs (*Melongena melongena*) in the mangrove woodlands. In contrast, Ortoire yielded mainly marine bivalves at home in a muddy, sandy-bottom habitat, especially Trigonal tivelas and Donax clams, rarely Ark shells (*Anadara* spp.). Clearly, these shells were collected on the beaches of Cocos Bay and Mayaro Bay. The occurrence of, for example, Thick lucinas (*Phacoides pectinatus*), West Indian Crown conchs, and Caribbean oysters indicates that the mangroves along the lower reaches of the Ortoire River were also exploited. Interestingly, the stratified character of both the St. John and Ortoire shell middens suggests that the shellfish collecting strategies of the Archaic Indians at these sites were highly focused, targeting particular species. This is confirmed by Harris's analysis of the Banwari Trace site (Harris 1973, 1976).

Ceramic Age Finds

The archaeozoological finds from the Saladoid–Barrancoid settlement sites of Cedros, Palo Seco, Quinam, and Erin, which were analyzed by Wing (1962) and Wing and Reitz (1982), provide information on the hunting, fishing, and collecting strategies of the horticultural Amerindians on Trinidad's south coast during Ceramic times. Recent investigations by Grouard (1998) of the archaeozoological materials encountered at the Ceramic settlement site of Manzanilla 1 on Trinidad's eastern shore can be used to supplement the picture arising from Wing's analysis.

Unfortunately, the samples of animal and fish bones collected by Rouse in 1946 were assumed to characterize the sites where he worked throughout their entire existence. There was no allowance for local alterations in subsistence strategies by examining the food remains according to stratigraphic position. As the archaeozoological materials recovered from the Saladoid–Barrancoid and Arauquinoid middens at Quinam were not kept separated, it is impossible to decide whether the Early and Late Ceramic hunting and fishing faunas at this site differed in the particular species targeted. No attempt was made to investigate the faunal remains of the Bontour site and the confused stratification of St. Joseph 2 renders it impossible to assign the (small) animal bone sample collected by Rouse and Goggin to either the site's prehistoric or protohistoric periods of occupation. Also note that the inference that swidden (shifting) cultivation with bitter cassava as the principal crop was practiced at Trinidad's Ceramic sites is based primarily on the presence of specialized pottery and stone artifacts at these sites (for example, griddles and tiny stone chips, used as "teeth" in cassava graters). Essentially, the subsistence strategies of Ceramic times attempted to arrive at a dietary balance between the input of carbohydrates (provided by growing starchy root crops, notably cassava and sweet potatoes, along with collecting wild plant foods) and the resources of proteins and fats secured by hunting, fishing, and the gathering of invertebrates such as crabs, shellfish, and echinoderms.

Current evidence suggests that the Ceramic age Amerindians of southwest Trinidad emphasized hunting and inshore–estuarine fishing. The Cedros, Palo Seco, Erin, and Quinam sites are situated in the vicinity of evergreen or semi-evergreen seasonal forest and most hunting apparently took place in the open forest environment, along small streams and near old and new swidden plots. Terrestrial game accounted for most of the vertebrate remains at Cedros (77.0%), Palo Seco (60.5%), Erin (64.6%), and Quinam (68.5%). Because no fine-mesh screens were used during excavation, fish remains are possibly underrepresented at these sites, suggested also by the somewhat lower figure obtained for exploitation of the terrestrial habitat (44.4%) at the contemporaneous Manzanilla 1 site (Boomert 2000, table 46). Red brocket, collared peccary, agouti, paca, nine-banded armadillo, and black-eared opossum were the six terrestrial mammal species preferred as game animals at the four southwest coast sites. Red brocket is the animal most frequently represented at these sites, on average accounting for 29.1% of the terrestrial mammals hunted (Boomert 2000, table 43). Sparse remains of terrestrial mammals such as red howler monkeys, three-toed anteaters, squirrels, spiny rats, prehensile-tailed porcupines, crab-eating raccoons, ocelots, and tapirs indicate that these were occasionally hunted as well. The same applies to aquatic mammals such as water rats, otters, and manatees. Remains of cachalot whales, typically offshore pelagic mammals, probably derive from animals that had washed ashore. Birds, amphibians, and reptiles were rarely used for food, although sea turtles are a notable exception. Fishing took place primarily in the inshore–estuarine habitat, just as in Archaic times. Sea catfishes and jacks were the fish families primarily targeted, according to the fish bones encountered at the four south coast sites. Other fishes caught in the inshore–estuarine environment included sharks, rays and skates, ladyfishes, snooks, spadefishes, mullets, tarpons, and stingrays. There were only a few examples of fishes from near the offshore banks and reefs (snappers, parrotfishes, and groupers).

The collecting patterns of the Ceramic age Indians who lived at the four settlement sites on Trinidad's south coast examined during the 1946 excavations—all shell midden deposits—clearly reflect the local habitats. The shell refuse at Cedros, Erin, Palo Seco, and Quinam suggests that a few particular mollusk species were targeted and other species were collected only as people came across them when searching for the preferred species. Each of the four south coast sites yielded tens, perhaps hundreds, of thousands of almost exclusively marine and estuarine mollusks. All the sites show similar patterns of shellfish collecting and preferences. Donax clams and Trigonal tivelas are the two shell species most frequently represented.

Cedros is characterized by predominantly Donax clams, less by Trigonal tivelas and Caribbean oysters (known locally as "tree oysters" because they live in colonies attached to the prop roots of red mangrove trees). The Cedros site also yielded a few West Indian Crown conchs, equally at home in a mangrove environment. Both species were undoubtedly collected in the extended mangrove woodlands east of the site. At Erin, Trigonal tivelas and Donax clams are almost equally represented. There are fewer of the mangrove species such as Caribbean oysters, West Indian Crown conchs, and nerites (Neritidae). Quinam yielded mostly Trigonal tivelas and fewer Donax clams, in addition to Caribbean oysters, West Indian Crown conchs, and Tulip mussels (*Modiolus americanus*). A few terrestrial snails of the genus *Strophocheilus* were also encountered at both Cedros and Quinam. Clearly, two major marine–estuarine habitat zones were exploited for shellfish at these three sites: the intertidal zone of the sand beaches and the mangrove woodlands stretching along the shoreline of the Columbus Channel, notably the Blanquizales Lagoon and the mouths of rivers such as the Erin River and the Quinam River.

Shellfish preferences were not any different at Palo Seco, but there was a major change in the predominant mollusk species gathered during the existence of this site. Palo Seco yielded at least twenty-six shell species (Bullbrook 1953:23–26, 68, 73, 1960:31). The lower stratum of the site is dominated by Trigonal tivelas, which in terms of minimum number of individuals represent 45.0% of the total shell species from this zone. Striate Donax clams and nerites range second and third, 22.1% and 16.4%, respectively. Other species collected include Rock shells (*Thais* spp.), known locally in Trinidad as "swamp conchs" (15.6%), and a few muddy-sand pelecypods such as Variable nassas (*Nassarius albus*) and Morocco naticas (*Natica marochiensis*). Moreover, Rouse's excavations yielded some Caribbean oysters, West Indian Crown conchs, and Tulip mussels, along with a terrestrial snail of the genus *Strophocheilus*. In the upper stratum of Palo Seco Striate Donax clams (87.4%) predominate and there is a dramatic decrease of Trigonal tivelas (5.5%). Whether the latter was because of overharvesting cannot be ascertained, although interestingly Trigonal tivelas are rare in the waters of Trinidad's south coast. Only "swamp conchs" are found in reasonable quantities (4.8%) in the upper stratum of Palo Seco; other shell species of marine–estuarine and mangrove habitats are rare. Consequently, the general pattern of the gathering habits of the Ceramic age Amerindians of Trinidad's south coast is that of a population emphasizing the collecting of intertidal beach pelecypods such as Donax clams and Trigonal tivelas, but also taking mangrove species such as Caribbean oysters, Tulip mussels, West Indian Crown conchs, and nerites.

The evidence does not suggest that there were major subsistence changes among the Trinidad Indians during the Ceramic age. Unfortunately, little information is available on the adaptive strategies of the Arauquinoid (Guayabitan) peoples of Late Ceramic times. Practically no animal bone was found at Bontour (Rouse Papers, Bontour field notes) and the collection recovered from St. Joseph 2 (only bone fragments of a porcupine, agoutis, a paca, and red brocket [Wing 1962:45, 49]) is too small to allow far-reaching conclusions. The mollusks encountered at Bontour and St. Joseph 2 reflect the environmental situation of these two sites. Bontour yielded predominantly Blood arks (*Anadara ovalis*) and small amounts of

Caribbean oysters, West Indian Crown conchs, Fighting conchs (*Strombus pugilis*), Tiger lucinas, and Venus shells (*Antigona* sp.). Other shell species are represented only by a few specimens, for example, Donax clams, Olive shells (*Oliva* sp.), Variegated turret shells (*Turritella variegata*), and True tulips (*Fasciolaria tulipa*). All are either at home in muddy, sandy-bottom shallow-water habitats or adapted to estuarine mangrove areas. Clearly, the shells were collected on the shore of the Gulf of Paria and its fringing mangroves, close to the Bontour settlement. At the Harris Promenade/High Street site, which is apparently the inland extension of Bontour, Caribbean oysters predominate whereas Tiger lucinas, Blood arks, and True Tulips are less abundant. St. Joseph 2 yielded mainly West Indian Crown conchs, Tiger lucinas, and various freshwater snails (Boomert 1985). Apart from the latter, the collecting areas of the Indians living at this site resemble those of Bontour. In contrast, the contemporary Guayabitan inhabitants of Palo Seco East typically collected Donax clams and Trigonal tivelas, the same species favored by the Early Ceramic Indians on the south coast. This suggests that the local habitat, not cultural differences, exclusively determined the mollusk species that were collected throughout the Ceramic age.

Historic Age Finds

The archaeozoological remains found at Mayo yield some insight into the subsistence strategies of the Amerindians living at a Spanish mission site in Trinidad during the eighteenth century. The missions were intended to be economically self-sufficient, producing and selling crops such as cocoa and rice for export and growing food crops for themselves and other missions in need. The Indians were bound to work two days a week on the communal lands of the mission; they were allowed to cultivate their own gardens for four days, probably practicing a modified system of shifting cultivation (Newson 1976:167; Borde 1982, 2:53–54). After the failure of the cocoa crop in 1727, the mission Indians were allowed to pay their tribute in cassava, maize, and tobacco. A 1777 inventory indicates that at the time the mission grounds were cultivated entirely with maize and plantains. A few horses and mares were probably kept for transport and chickens, pigs, and goats raised for consumption (Dorta 1967:154; Newson 1976:168). Hunting and fishing were important subsistence activities. According to a Spanish visitor, in the late 1770s the Amerindians of the four missions in the Naparima area of west–central Trinidad— Guayria (present San Fernando), Sabana Grande (present Princes Town), Savaneta, and Mayo—mainly hunted for peccaries and armadillos, assisted by many dogs raised especially for hunting, and caught turtles and mullets in the channels of the Orinoco Delta. Fish were dried and kept for a long time. Collecting palm nuts was an additional subsistence activity ("Concerning the Islands..." [1780]).

The animal bones found at Mayo confirm the historical records. The excavations by John and Rita Goggin in 1953 yielded mostly remains of hunted fauna such as collared peccaries and nine-banded armadillos, fewer red howler monkeys, agoutis, and red brocket, and some porcupines and pacas. In addition, bones of a dog were found (Wing 1962:44-46). The archaeozoological materials recovered by Anderson-Córdova at Mayo in 1978 include predominantly turtles (*Dermochelys* sp.) and some remains of catfish (Ariidae), red howler monkeys, black-eared opossums, along with unidentified rodents and a carnivore. The remnants of a small ruminant (Ungulate) were identified (Victoria Jones, pers. comm. to Boomert 1984). The presence of large quantities of shells of marine–estuarine species at the Mayo site is suggestive of the collecting of mollusks on the shore of the Gulf of Paria, some 6 km to the west, as the crow flies. The Goggins encountered principally Tiger lucinas, fewer Caribbean oysters and West Indian Crown conchs, and some West Indian murex (*Murex brevifrons*), as well as a few freshwater snails, including River conchs and Ramshorn snails (*Marissa cornuarietus*). The shell remains (366 minimum number of individuals) found by Anderson-Córdova in 1978 were investigated by Victoria Jones (pers. comm. to Boomert 1984). Tiger lucinas and West Indian oysters are predominant (42.3% and 37.7%, respectively), with West Indian Crown conchs ranking third (11.5%). In addition, a few freshwater snails— River conchs (6.0%) and Ramshorn snails (1.4%)—and some terrestrial gastropods, notably *Eudolichotus distaria* (1.1%), were recovered. All this suggests that as late as the eighteenth century the Amerindians of Trinidad's Spanish mission villages followed subsistence strategies that in many respects still closely resembled those of their predecessors in pre-Columbian times.

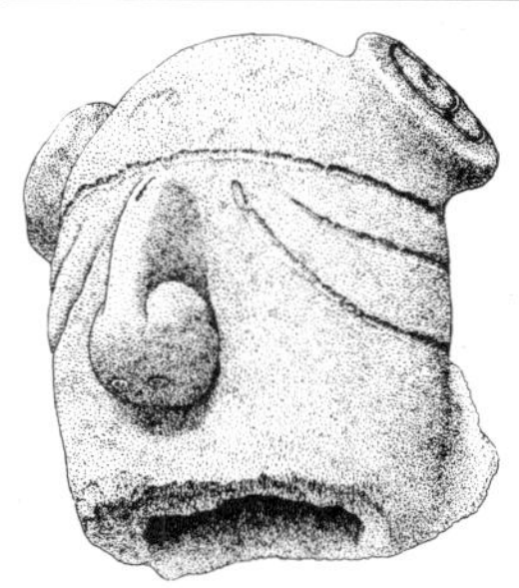

Conclusions

Irving Rouse's 1946 and 1953 excavations in Trinidad answered fully to his primary research objectives, establishing a local chrono-cultural sequence that could be used to bridge the developmental frameworks then available for the West Indies on the one hand, and the Orinoco Valley and coastal Venezuela on the other hand. In the following years Rouse was to devote his energies to refining the chronology of the cultural succession especially for the latter area. In the 1960s, this resulted in the first version of his Caribbean-wide chrono-cultural taxonomy, which would be revised several times during the next three decades (Rouse 1964, 1978, 1986, fig. 23, 1992, figs. 8–9, 14; Rouse et al. 1995; Rouse and Faber-Morse 1999, fig. 26). Clearly, Rouse's Trinidad research aims were distinctly classificatory–historical, since he theorized that archaeological data accumulation in the Caribbean had insufficiently advanced to allow inferences beyond the reconstruction of cultural traditions and chronological sequences. In the past half century or so Rouse's chrono-cultural framework has surely proven its usefulness for structuring the prehistoric sequence of the Caribbean, although its principally phylogenetic character and the implicit assumption of cultural homogeneity of its units have recently been criticized (Keegan 2001; Curet 2004). Rouse's charcoal samples from Trinidad's Ortoire site submitted for dating to the then newly founded Geochronometric Laboratory at Yale University resulted in the first radiocarbon measurements of archaeological materials from the Caribbean (Preston et al. 1955). His Trinidad investigations were also the starting point for archaeozoological research in the region, as Wing's study of the animal and fish bone materials recovered from Rouse's excavations was the beginning of the detailed analysis of prehistoric subsistence patterns and modes of life in the Caribbean (Wing 1962), thus signaling the beginning of a shift in archaeological research from an exclusively culture–historical interpretation toward contextual reconstruction and understanding.

Trinidad's Prehistoric Cultural Chronology

More than half a century after Rouse's research in Trinidad, our insight in the chrono-cultural sequence of the island still conforms in broad lines essentially to the preliminary framework he envisaged. Subsequent work by Harris (1976, 1978, 1991a) and Boomert (1985, 2000) has elaborated and modified Rouse's prehistoric chronology and cultural classification without significantly altering it. Accordingly, most of the successive Lithic, Archaic, and Ceramic cultural units distinguished in Trinidad's pre-Columbian past have been defined following Rouse's guidelines of cultural taxonomy and named after sites he excavated. The most spectacular progress in our knowledge has been on the Lithic and Archaic occupation of the island, mainly as a result of the excavations by Harris at the sites of Banwari Trace, St. John, and Poonah Road from 1969 through 1972 (those at Banwari Trace were resumed in 2005 and 2006). After Rouse's short stay in Trinidad in 1969 to collect samples for radiocarbon dating at Cedros and Palo Seco, Harris and members of the Trinidad and Tobago Historical Society (South Section) conducted surveys and test excavations of various Ceramic age sites on the island. They were joined by Boomert in the 1980s, who was able to realize a modest archaeological research center at the Trinidad campus of the University of the West Indies. In the past decade Harris has continued this excavation program assisted by members of the Trinidad and Tobago Archaeological Volunteers. This effort has led to a better understanding of the Saladoid through Mayoid occupations of the island, emphasizing pre-Columbian material culture, adaptive strategies, regional interaction patterns, and world view, by examining sites such as Atagual, Blanchisseuse, Guayaguayare, Icacos, San Fernando–Carib Street, and St. Catherine's. A major open-area excavation project was carried out by a Dutch team between 1997 and 2007 under the auspices of the National Archaeological Committee of Trinidad and Tobago at the Saladoid–Arauquinoid site of Manzanilla 1 on Trinidad's east coast (e.g., Nieweg and Dorst 2001; Dorst 2004, 2006, 2008). Archaeological research at the sites examined by Rouse on the southwest littoral was not undertaken again until 2005, when Boomert conducted test excavations at Cedros.

Lithic through Archaic Age Occupations

The presence of nomadic groups of Lithic hunters and foragers in Trinidad about 8000 cal BC—thus predating its separation from the South American mainland—is recognized only because of the individual find of a Joboid spearhead at Biche in east-central Trinidad (Harris 1991b). Fortunately, the Archaic occupation of the island can be reconstructed to a much higher resolution. At present a more or less continuous succession of culturally affiliated Archaic assemblages can be distinguished on the island, starting around 6000 cal BC. Dating, habitat correlations, material culture, and food procurement strategies of most of these complexes, grouped as members of the Caribbean-wide Ortoiroid series, have been estimated to a considerable degree. It has been possible to show that the Archaic inhabitants of Trinidad followed highly diversified subsistence strategies aimed at exploiting a wide range of environmental niches using generalized though highly focused procedures. Seafaring and navigating with large dugouts was an integral part of Trinidad's Archaic cultural heritage and the first movement of Amerindians into the Lesser Antilles by Ortoiroid Indians from Trinidad and the east Venezuelan littoral took place by about 5000 to 4000 cal BC. Individual finds of grooved axe and adze heads on the island can be interpreted partially as special activity areas for felling trees used to make canoes.

Clearly, Trinidad's Archaic communities kept in close contact with the contemporary Indians of the Paria Peninsula, the west coast of the Gulf of Paria, Tobago, and the Windward Islands. Most likely they followed coastal routes for transport and communication, keeping in sight of land and navigating by observing the celestial bodies and using dead reckoning on the open sea, just as their descendants did in protohistoric times. Did the Margarita–Paria–Trinidad area function as a kind of "voyaging nursery" comparable to the one in Melanesia in which, as Irwin (1992:24) hypothesizes, maritime technology was able to develop gradually in the initial stages of the prehistoric colonization of the Pacific? Island-hopping and paddling between the mainland and the islands (for instance, between Margarita and Cumaná) is easy in the Margarita–Paria–Trinidad "corridor" and can be done without losing sight of the mainland. Also, this region, at the southern fringe of the Caribbean hurricane belt, is seldom struck by destructive tropical storms. Finally, as in the Bismarck Archipelago, the seasonal and predictable changes in wind direction may have encouraged the early voyagers to experiment with round-trip expeditions exploring and exploiting the intervisible islands and coastal stretches.

Saladoid Colonization and Development

There is no evidence that the Ortoiroid series of Trinidad survived as late as the first Ceramic age settlers on the island, but this seems to have been different on the islands offshore of the east Venezuelan littoral. Here clearly Saladoid "trade" pottery has been found in association with the lithic and shell industry of the Late Archaic Punta Gorda complex of the Manicuaran subseries. The Warao Indians of the Orinoco Delta today, who until recently subsisted exclusively on hunting, fishing, and food collecting, may be the descendants of the Ortoiroid Indians of Archaic times. Whether the Manicuaran Indians and the first Saladoid immigrants of the region, arriving under Barrancoid pressure from the Lower Orinoco Valley, established a kind of symbiotic subsistence pattern, exchanging horticultural products for fish and game meat, is difficult to determine. However, there can be hardly any doubt that the Saladoid settlers learned about the existence of the southern Windward Islands and the necessary maritime technology and navigational requirements to reach these islands through interaction with the local Manicuaran Indians, who seem to have been acquainted with at least Grenada and the Testigos (Boomert 2000:83, 88, 221–222). Because north of Grenada the islands of the Lesser Antilles are almost permanently intervisible, traveling as far north as the Anegada Passage, which separates the Leeward Islands from the Virgin Islands and Puerto Rico, is easy (Keegan 2004). Recently, it has been stressed that the Saladoid peoples moving into the West Indies likely entered a peaceful relationship with the sparse Archaic Indians occupying the archipelago, co-habiting with them and gradually absorbing their communities (Curet 2005:67–70). This would replace the previously held scenario of a principally inimical relationship between both populations, at least in the Antilles.

The last few centuries before the beginning of our era saw the gradual disappearance of the Archaic communities in the region, the postulated partial withdrawal of the Ortoiroid Indians into the Orinoco Delta, and their rapid assimilation by the peoples of the Saladoid series. Subsisting on a combination of rootcrop horticulture, hunting, fishing, and the gathering of animal and plant foods, the Saladoid colonists were able to adapt very successfully to their local environment, with enough demographic growth ultimately to allow dense occupation of Trinidad and the east and central Venezuelan coastal zones (Boomert 2003). Attracted by the favorable conditions of the Antillean archipelago, adventurous young men in the local communities of the mainland and Trinidad initiated the Saladoid expansion into the Windward Islands and beyond. Apparently, the primary movement was rapid, perhaps because few local Archaic settlers were encountered, especially in the Windward Islands. Such a fast initial spread in the Antilles is incompatible with one allowing for a step-by-step colonization of each island due to demographic growth. In fact, the scenario of a multiple series of fast initial migrations forming a direct leap forward along the Lesser Antillean island chain to the Leeward Islands, the Virgin Islands, and Puerto Rico has been envisaged, followed by various return movements finally leading to the settling of all the islands previously passed by (Keegan 1995, 2004). Be this as it may, during the first centuries of Antillean settlement, Saladoid communication and interaction across the West Indian archipelago remained intensive, as suggested by the widespread occurrence of exotic exchange objects, notably ornaments made of semi-precious rock materials, some of which may have reached the West Indies from the South American mainland and Trinidad (Faber-Morse 1989:34).

Sustained interaction is shown also by the remarkable uniformity of the pottery typifying the earliest Saladoid communities throughout the West Indies (Allaire 2003). Well-finished, small to medium-sized bowls and jars with a variety of bi-chrome and polychrome painted, incised, and modeled designs are diagnostic. Rouse (1986:134) coined the term Cedrosan, after Trinidad's Cedros site, to define the Saladoid ceramics of the West Indies. The close relationship between the Cedrosan assemblages of Trinidad and the Windward Islands is shown by the resemblances between the pottery repertoire of the Cedros complex and that of Early Vivé in Martinique (Bérard 2004:151–152, 190). Note that to date there is no evidence of the presence of the contemporaneous Huecoid series in either Trinidad or the east Venezuelan coastal area. The local evolution of Saladoid ceramics on the island reflects regional stylistic developments during Early Palo Seco times. As we have seen, ZIC designs disappeared rapidly on Trinidad's southern shore after the Cedros complex, whereas elsewhere in the Lesser Antilles they remained in fashion (e.g. Rouse and Faber-Morse 1999, fig. 12). Moreover, on the north coast a specific type of thin-line incised, or rather engraved, designs now developed out of Cedros incision. Characterized by single or multiple parallel wavy lines and semicircles, spirals, and stepped lines, it closely resembles the incised motifs of the pottery of the Mount Irvine complex of Tobago (Boomert 2000:171–179), the Pearls Inner Rim Incised and Grande Anse Interior Incised types of the Windward Islands, and some of the incised motifs on the Río Guapo ceramics of the central Venezuelan coast. This suggests the development of some form of internal stylistic differentiation on the island, effected by the distinct interaction patterns of the Saladoid communities north and south of the Northern Range, which apparently acted as a natural barrier to Trinidad's prehistoric intra-island communication.

Saladoid Religion and Social Order

The profusion of anthropozoomorphic head lugs ornamenting the Cedros to Palo Seco ceramics of Trinidad adequately reflects the animistic nature of Saladoid religion and the profound Amerindian belief in human–animal transformations. Indeed, many vessels are really zoomorphic effigies, representing "mythic transforms" of creatures, which also act as the major natural symbols in South American mythology. The Saladoid zoomorphic *adornos* include highly conventionalized representations of birds and animals such as monkeys, turtles, armadillos, bats, frogs, dogs, rodents, felines, caimans, and others. Many of these head lugs show a form of pictorial dualism, that is, the occurrence of divergent images when seen from contrasting points of view (Keegan and Byrne 2001; Reid 2004; Moravetz 2005:79–80). The wish to invoke the direct approval of the spirit world during shamanic rituals and in other ceremonial contexts must have impelled the use of vessels provided with biomorphic head lugs. Effigy vessels used during shamanic ceremonies may have been the temporary repositories of the shaman's guardian spirits when summoned for advice and assistance. The varied iconography shown by Cedrosan pottery indicates the high value placed by the Ceramic age Indians on communication with the spirit world to ensure health, fertility, social order, and group survival. As noted in Chapter 3, the shamanic ingestion of hallucinogenic drugs to induce an ecstatic–visionary trance is indicated by the common occurrence of double-spouted "nostril bowls," used to pour tobacco or pepper juice into the nose, at the Saladoid sites of Trinidad. Other Cedrosan shamanic aids found on the island include vessels provided with hollow biomorphic head lugs containing small clay pellets or tiny pebbles, which obviously functioned as rattling devices during curing ceremonies, and ceremonial pottery cylinders or incense burners (Rouse and Faber-Morse 1999:38). Remarkably, only one stone treepointer, belonging to the most ubiquitous shamanic paraphernalia elsewhere in the Caribbean, has been found in Trinidad. It may be an exchange item from Tobago or the Windward Islands.

The variety of forms, the quality of manufacture, and the intricate ornamentation of the Saladoid pottery suggest that its ceremonial component was principally destined for the public display of food and other properties as an expression of status, rank, or kinship affiliation. Such displays could have been part of ceremonies involving competitive demonstrations of wealth accompanied by gift giving or even property destruction. The ornamented fine ware may have been used primarily for serving food during ceremonial feasts, held for the extended following of local headmen, "great men" who, through their personal qualities, were able to dominate in war and exchange, attracting large followings through gift giving. Feasts of this kind would have been held on such occasions as initiation rituals, marriages, and the burials of high-ranking persons. If so, although a principally egalitarian society, fluctuating status differences among its male adults can be postulated for the Saladoid social order. These may have been expressed during public ceremonies of competitive emulation leading to the deliberate deposition or "ritual killing" of valuables in displays of conspicuous wealth (Boomert 2000:382, 392–394, 441, 2001). Such exotic-looking, highly valued, and rare objects, predominantly bodily ornaments, were manufactured and exchanged in large numbers in the West Indies throughout Saladoid times. Many of these items are beads and small zoomorphic pendants made of nonlocal rocks and minerals, including semiprecious stones (e.g., Faber-Morse 1989:34). Clearly, they represent the products of highly expert village artisans and semispecialists, who may have combined this craftmanship with shamanic activities. Saladoid-associated stone accoutrements of this kind found in south and central Trinidad include turquoise, chlorite, and serpentinite beads; unworked pieces of amethyst may have been intended for the manufacture of such ornaments. Finally, the modified or unworked valves of nacreous freshwater mussels, encountered in Saladoid context at the Cedros, Palo Seco, and Erin sites, obviously were highly desired items of exchange throughout the Lesser Antilles, since they have been found as far north as the Leeward and Virgin Islands (Serrand 2001).

Saladoid–Barrancoid Interaction

The gradual adoption of Barrancoid ceramic modes by the Palo Seco potters of Trinidad reflects the growing interaction between the Saladoid communities of the island and the Los Barrancos (Classic Barrancas) complex that had been developing on the Lower Orinoco since the first centuries of our era (Sanoja and Vargas 1983; Gassón 2002). This process is repeated in the eastern coastal zone of Venezuela and on the middle reaches of the Orinoco. It is suggestive of frequent contacts formalized by ceremonial exchanges in a localized interaction sphere (Rouse 1983; Boomert 2000: 239–240, 442–444). The inorganic valuables that seem to have circulated through the system were predominantly pottery, along with small amounts of bone and stone ornaments. The ceramics exchanged included Barrancoid vessels showing intricate forms and decorative motifs such as elaborately modeled-incised biomorphic head lugs, suggesting representation of spirit-related messages and strongly shamanic associations. These Barrancoid exchange items reached as far as Tobago, where they were used as mortuary gifts, indicating the high prestige these exotic vessels had among the local Saladoid population. As the Barrancoid peoples apparently were the dominant partners in the Saladoid–Barrancoid exchange relationship, it must be assumed that they were primarily responsible for carrying out the trade expeditions radiating from the Lower Orinoco Valley. These patterns of island–mainland interaction must have promoted the exchange of information and nonmaterial goods in the form of myths, tales, songs, dances, and knowledge, leading to the conclusion of political alliances based on kinship ties and ritual services between the communities involved. The growing intensification of local interaction in the region is also reflected by the presence of many stone artifacts, especially axe heads, at the Saladoid sites on Trinidad, apparently imported from Tobago, the Windward Islands, and the Paria Peninsula.

The regular and institutionalized interaction between the Saladoid and Barrancoid communities of the region culminated in the establishment of a local Barrancoid ceramic complex, Erin, in Trinidad. While found as a "trade" ware at the contemporaneous Saladoid sites throughout the southern and central portions of the island, Erin pottery is exclusively represented in substantial amounts on Trinidad's southwestern littoral, notably at the Chagonary, Erin, Quinam, and La Lune 1 multicomponent settlement sites. Moreover, it is consistently encountered in close association with (Late) Palo Seco ceramics, suggesting largely simultaneous manufacture and use of both wares. All this suggests that people of Barrancoid cultural affiliation originating in the Lower Orinoco Valley settled on the island to live side-by-side, and no doubt intermarry, with the Saladoid inhabitants of the southwest coast. A similar scenario can be hypothesized for the establishment of the Erin-like Chuare complex in the eastern Venezuelan coastal zone.

Thus far the onset of Erin settlement in Trinidad has been placed at a date of about cal AD 350, after a few centuries of exclusively Barrancoid–Saladoid exchange. However, our examination of the stratigraphic sequence shown by the pottery of the Palo Seco, Erin, and Bontour complexes, especially at the Palo Seco, Erin, and Quinam sites, indicates that apart from vessels tempered with particles of deliberately crushed quartz in typical Los Barrancos fashion, a considerable portion of the Erin ceramics at these sites are characterized by pottery tempered with fine quartz sand, sponge spicules (*cauixí*), or a combination of both nonplastics (Petersen 2004; Faber-Morse 2009). They were decorated with incised and modeled designs showing a quite loose, indeed sometimes distinctly haphazard and sloppy, execution clearly reflecting a definite slackening of the strict standards of decoration regulating "classic" Barrancoid pottery. This suggests that most Erin pottery in south Trinidad is affiliated with the Guarguapo (Post-classical Barrancas) complex of the Lower Orinoco rather than to Los Barrancos. Guarguapo reflects the onset of Arauquinoid influence on the Barrancoid ceramics of the region (Sanoja and Vargas 1983; Gassón 2002). All this indicates that the independent establishment of the Erin complex in south Trinidad may have taken place at a somewhat later date than previously assumed, perhaps about cal AD 500.

Establishment of the Arauquinoid Series

The period from cal AD 500 to 700 is one of exceptional dynamism in the Orinoco Basin. It is characterized by the movement of Indians of the Arauquinoid series from the middle to the lower reaches of the river, the subsequent establishment of the Arauquinoid Macapaima complex on the Lower Orinoco, and the contemporaneous stylistic transformation of the Barrancoid pottery of the Los Barrancos (Classic Barrancas) complex to that of the Guarguapo (Post-classical Barrancas) complex by the adoption of Arauquinoid modes of manufacture, such as *cauixi* temper and the gradual disappearance of typically Barrancoid decorative elements. We hypothesize that the settlement of Guarguapo people in south Trinidad was connected with the downstream movement of the Arauquinoid Indians in the Orinoco Valley, somewhat comparable to the postulated way the Saladoid peoples were pushed to the Venezuelan coastal zone and Trinidad by the Indians of the Barrancoid series. Simultaneously with the onset of the Erin complex in the southwestern portion of Trinidad, another Arauquinoid-influenced Barrancoid ceramic complex developed at the Guayaguayare and St. Catherine's sites in the southeasternmost part of the island. This St. Catherine's complex is closely related to the Apostaderan subseries of the Orinoco Delta and the coastal zone of Guyana, suggesting dense interaction across the Columbus Channel in this area. Eventually the Arauquinoid domination radiating from the Lower Orinoco would stretch as far as the east Venezuelan coastal zone, here giving rise to the Macapaima-related Guayabita complex, and Trinidad where Erin, (Late) Palo Seco, and St. Catherine's would be replaced by the Bontour complex. Macapaima, Guayabita, and Bontour are sufficiently alike to justify incorporation of these three pottery assemblages into a Guayabitan subseries of the Arauquinoid series (Boomert 2003), suggesting cultural unification of the entire region of the previous Saladoid–Barrancoid interaction sphere by about cal AD 650–700.

The Bontour complex is typified by shell-tempered open bowls and simple jars with inflected contours sometimes showing punctated appliqué fillets at the neck bases. Decoration is reduced to an absolute minimum. It is restricted to sparse punctated and incised or gouged motifs and shows an absolute break with Saladoid–Barrancoid times in the almost complete disappearance of biomorphic head lugs. This decline in the quality of manufacture and decoration of the local Trinidadian ceramics can be explained only by assuming that pottery had lost many, if not all, of its ritual messages to other items of material culture. It resembles the situation that characterized the Lower Orinoco region from the onset of the Arauquinoid series and may indicate that ceramics were much less imbued by ceremonialism than during Saladoid–Barrancoid times. Perhaps shamanic eschatological expression now manifested itself on artifacts made of perishable materials.

All this suggests that the transition from the Saladoid and Barrancoid to the Arauquinoid series in Trinidad and elsewhere meant a genuine cultural break. Of course, this does not mean that the original population of the island was completely replaced by newcomers. To the contrary, the gradual succession of Saladoid and Barrancoid ceramics by the Bontour complex suggests a steady development and cultural transition involving primarily a locally seated population. It would explain, for instance, the apparent continuity of the local modes of (Late) Palo Seco ceramic technology such as siltstone pottery temper well into Arauquinoid times at the

Manzanilla 1 site (Dorst 2004, 2006, 2008). This settlement site is exemplary for the many Saladoid and Saladoid–Barrancoid sites that continued to be inhabited during the Bontour complex. The excavations at Manzanilla 1 have shown that, in the Bontour period, villages were characterized by circular house structures surrounded by shell midden deposits where inhumation burials were placed (Dorst 2006:17). The many more sites, if compared to Saladoid–Barrancoid times, suggests an increasing population density throughout the island.

Communication and interaction with the mainland continued throughout the Bontour episode, as shown by the presence of particular exchange objects from the Lower Orinoco, notably griddles and ceramic roller stamps, both tempered with sponge spicules, in Trinidad's Bontour settlements. In the Orinoco Basin such roller stamps, most likely used for body painting, were prestigious items of ceremonial exchange throughout Arauquinoid times. Conversely, shell-tempered Bontour pottery has been recovered from the Lower Orinoco Valley. In addition, typically Apostaderan keeled vessels originating in the coastal zone of Guyana reached as far as the Erin and Icacos settlement sites of Trinidad's southwestern littoral. Intra-island interaction is suggested by the recovery of pottery tempered with river sand containing micaschist particles, typical of the Northern Range, at the Bontour sites of the southern part of the island. Finally, contacts with Tobago and the Windward Islands are indicated by the presence of Bontour ceramics on Tobago, the Testigos, and Carriacou (Grenadines). On Tobago, Bontour pottery has been found at the sites of Golden Grove and Sandy Point, associated with the ceramics of the Golden Grove complex, a local Troumassan Troumassoid assemblage (Boomert 2005, 2009). Interestingly, both a specific vessel shape and several modes of incised or gouged and modeled decoration, especially rim modifications, typical of Golden Grove ceramics, seem to duplicate Bontour examples. They find their counterparts in the Troumassan pottery of the Windward Islands. The latter designs have often been seen as local simplifications of decorative modes dating back to Saladoid times, but current evidence suggests that the transformation from Saladoid to Troumassoid in the Windward Islands may have taken place under the stimulus of the inception of the Arauquinoid series in Trinidad in a way quite similar to that characterizing the latter island's development from Saladoid–Barrancoid to Arauquinoid (Boomert 1985, 2005, 2009, 2010). Most likely Tobago played a mediating role between Bontour and Troumassan ceramics, since here influences from the Windward Islands and Trinidad seem to converge.

Amerindian Society in the Historic Period

The final Amerindian ceramic tradition of Trinidad, the Mayoid series, is characteristic of the Amerindian–European contact period. It emerged perhaps shortly before Columbus passed by the island in AD 1498 on his third journey to the West Indies. Mayoid pottery may have been manufactured until the mid-eighteenth century. At some sites sparse amounts of Mayoid ceramics have been found associated with Bontour pottery, suggesting a certain period of overlap, but at a site such as Guayaguayare in the southeast there is a sharp stratigraphic break between both complexes. The exclusive use of *caraipé* as tempering material for Mayoid pottery and the strong resemblance of the characteristic Mayoid cooking jars with the "buck-pot" of the Arawak (Lokono) in the Guianas suggest a derivation of the Mayoid series from this region (Boomert 1985, 1986). The rare occurrence of a typically Koriabo–Cayo necked vessel in Mayoid context points in this same direction. While Mayoid ceramics were apparently manufactured by the Nepoio and Arawak (Lokono) Amerindians in Trinidad, it has been hypothesized that the Koriabo–Cayo pottery represents the predecessor of most present and subrecent Amerindian pottery traditions of the coastal zone of the Guianas and the Windward Islands, including those of the Palikur, Kalina, Arawak, and Island Carib ethnic groups (Boomert 2004). This suggests that most of the Amerindians in the region shared, generally speaking, one tradition of pottery manufacture in colonial times. Moreover, Mayoid ceramics have been encountered in Spanish context at St. Joseph, Trinidad's first Spanish capital (AD 1592), which remained the major center of Spanish occupation on the island well into the eighteenth century (Silver and Faber-Morse 2010). It can thus be assumed to have been the domestic pottery the Spanish and *mestizos* of St. Joseph used during colonial times, obtaining these vessels from the loyal Amerindians living in the *encomiendas* and, afterward, the missions of the island. As noted above, the paucity of *majolica* and the preponderance of Mayoid pottery in the Spanish capital of Trinidad reveal the poor living conditions of its inhabitants. It was not until the 1780s that the Amerindian population of the island was numerically surpassed by people of European and African descent (Silver and Faber-Morse 2010).

At the time of the European–Amerindian encounter, Trinidad formed a complex multiethnic and multilingual conglomerate of Amerindian groups of possibly varying sociopolitical complexity (Boomert 2009). Intra-island interaction was limited; communication typically took place by boat, skirting round the island, and this is the pattern that dominated transport from prehistoric times until the early twentieth century. Clearly, the Northern Range was the major obstacle to interaction across Trinidad's landmass, effectively isolating the north coast from the rest of the island. Extra-insular interaction was much more frequent and momentous. Because of its geographical position at the crossroads of waterways between the South American mainland and the Caribbean archipelago, Trinidad formed a central hub in the system of Amerindian exchange, communication, and diffusion of culture that developed in the region during prehistoric times, encompassing the Orinoco Valley, the Gulf of Paria, the Guiana coastal zone, and the southern islands of the Lesser Antilles. Trinidad's location between the mainland and the other islands is reflected by its prehistoric cultural sequence and the island's Amerindian cultural geography, which can be reconstructed from the pertinent documentary evidence dating back to the contact period. The fragmented character of its early historical Amerindian ethnicity, linguistics, and sociopolitical configuration is matched by the similarly disunited make-up of the Lower Orinoco Valley and the coastal zone of the mainland. The succession of the Ortoiroid, Saladoid, Barrancoid, Arauquinoid, and Mayoid series in Trinidad largely corresponds with the cultural sequence encountered in the Lower Orinoco valley and delta as well as the east Venezuelan coastal zone. This suggests that there was a continuous flow of information between the Amerindians of these areas throughout most of the Archaic and Ceramic ages. Trinidad and the adjacent parts of the South American mainland must have formed one extensive interaction sphere, tightly knit by ties of kinship, language, exchange, war, and culture. Apparently, developments taking place on the continent were quickly felt in Trinidad, and vice versa. All this shows that the island should not be conceived of as part of the Lesser Antilles, but as an extension of the mainland of South America—which in a physical sense indeed it once was.

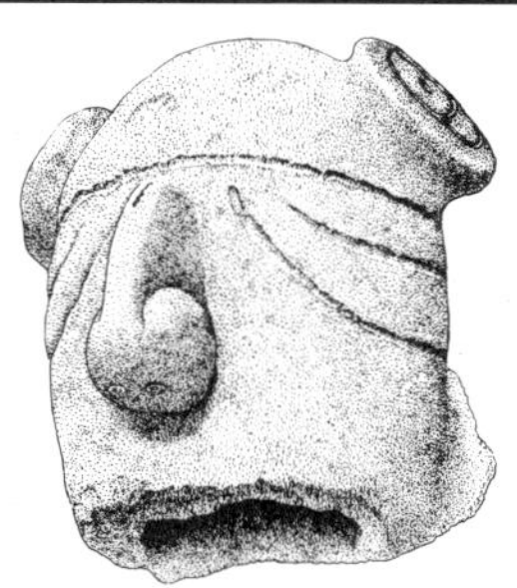

Geochemical Pottery Analysis

A. J. Daan Isendoorn and Arie Boomert

Our current knowledge of the provenience of raw materials and finished products of Amerindian pottery in the insular Caribbean is limited, whereas ethnohistoric evidence and finds of lithic or shell objects suggest that formerly such items and ceramics circulated widely in exchange networks throughout the archipelago. In recent years the study of clays, temper materials, and potsherds of prehistoric and protohistoric vessels from the Antilles to reconstruct these patterns of inter-island exchange got a fresh impetus from conventional archaeological methods, including workability tests, technological experiments, and microscopic fabric analysis combined with geochemical research such as X-ray fluorescence (XRF) spectrometry and ethnoarchaeological studies (Hofman et al. 2005, 2008). This Appendix summarizes the results of the XRF analysis of a collection of pre- and protohistoric potsherds from Trinidad curated in Division of Anthropology at the Yale Peabody Museum of Natural History (YPM), New Haven, Connecticut, USA, and at the Department of History of the University of the West Indies (UWI), St. Augustine, Trinidad. The analysis was carried out under the auspices of the long-term project "Mobility and Exchange: Dynamics of Material, Social, and Ideological Relations in the Pre-Columbian Caribbean," directed by Corinne L. Hofman, professor in the Faculty of Archaeology of Leiden University, The Netherlands, and financed by The Netherlands Organisation for Scientific Research (NWO), The Hague.

Materials and Methods

The collection of potsherds examined includes eighty-seven prehistoric and protohistoric pieces recovered from fifteen archaeological sites in Trinidad. The collection is curated in both the Yale Peabody Museum (fifty-nine potsherds from six sites) and the Department of History of the University of the West Indies, St. Augustine, Trinidad (twenty-eight potsherds from nine sites). The potsherds (selected by Boomert) encompass pieces that can be assigned to various pre- and protohistoric Amerindian cultural assemblages, including the Cedros and Palo Seco complexes of the Saladoid series, the Erin complex of the Barrancoid series, the Bontour complex of the Arauquinoid series, and the Mayo complex of the Mayoid series. Together they present an overview of the entire ceramic cultural sequence of

Amerindian Trinidad, ranging from the first centuries BC until the mid-eighteenth century AD. The Yale finds were recovered during the expeditions to Trinidad in 1946 and 1953 by Irving Rouse; the UWI pieces are from surveys and excavations made by Boomert in collaboration with Peter O'Brien Harris, in Trinidad, during the years from 1980 to 1988.

In addition to the potsherd samples, field work in Trinidad, carried out in 2006 by Daan Isendoorn in collaboration with Mathijs A. Booden, yielded twenty-four clay samples from source locations in the neighborhood of the archaeological sites represented. These clay samples were used for workability tests to evaluate the suitability of the clays for pottery manufacturing techniques such as coiling, molding, pinching, flattening, smoothening, and drying, as well as to test their firing and post-firing behaviors. Experimental work and ethnoarchaeological observations show that traditional potters on the Caribbean islands used similar techniques for earthenware manufacture (Hofman and Bright 2004; Hofman and Jacobs 2004). Stereomicroscopic fabric analysis was used for the identification of the mineral and nonmineral constituents in the clays and potsherds. The latter were cut with a diamond saw and treated with fine sandpaper to obtain a smooth fraction surface. Finally, the potsherds were re-fired in an oxidizing oven at 750 °C for thirty minutes. This removed the organic impurities in the fabric and improved the visibility of the inclusions and paste color.

The XRF analysis was carried out at the Faculty of Earth and Life Sciences of the Free University, Amsterdam, The Netherlands, in collaboration with Gareth R. Davies. XRF was performed to identify the chemical signatures of the raw clay materials and the Amerindian potsherds to trace their provenience. The element ratios in the clays and potsherds show several distinct, geographically related groups. By connecting the pottery to its presumed clay source, exotic potsherds can be identified within the sample. XRF determines the composition (geochemical signature) of a sample by calculating the exact number of the various elements. This method is useful for determining the provenience of pottery samples when combined with the results of the elemental compositional analysis of the clay samples. The clay samples were oven dried, then crushed in a jaw crusher, homogenized, and finally pulverized. The powdered samples were then pressed into pellets. These were loaded into the XRF spectrometer for a quantitative analysis of the elements. Using the same archaeometric techniques, the chemical signatures of the potsherds could be compared with those of the clays.

Results

The various geological formations of Trinidad are quite different in the quality and quantity of clay sources suitable for pottery manufacture. The island has five major physiographic units: three progressively lower chains of mountains aligned on an east–west axis (the Northern Range, the Central Range, and the Southern Range) and two intervening lowlands (the Northern Basin and the Southern Lowlands).

Clays were sampled from the three areas where the archaeological sites are located, from which the potsherds have been analyzed: (1) the Northern Range, (2) the San Fernando area of central–west Trinidad, and (3) the Cedros Peninsula in the southwest and the central portion of the island's south coast. The data obtained from the XRF analyses are presented as geochemical bivariate diagrams. First the degree to which the compositions of bedrocks, clays, and potsherds have been altered geochemically by weathering will be determined. Second, trace element ratios are used to distinguish the clay sources for the various potsherds.

The major formations in the Northern Range are generally capable of producing proper clays. The phyllites of the Maracas and Guayamare formations seem to be the main sources of suitable pottery clays. These clays look very similar and have a distinctly "dry" texture, probably resulting from their mica content. The highest areas of the Northern Range provide the richest clay sources. The small formations in the northeast, including the Sans Souci volcanics, are probably clay forming, but have not been visited because of their remote location. In the San Fernando area it was the extensive Chin Chin Formation that yielded many excellent clay soils and bedrock clays. The latter especially are very pure and plastic when wet. The argillite of Naparima Hill did not yield any clays, despite the promising lithological name. Farther south, the Lower Cruse Formation, which crops out over a wide area, provides several clayey soils. Finally, along the Cedros Peninsula and the south coast as far as Moruga, a sequence of poorly consolidated sediments, from shaley clays to sandstones, is exposed. The best and purest clays were collected in direct exposures of the bedrock, especially in the area of the Erin Formation. The bedrock clays are usually best in quality if they derive from a surface that has been exposed for a while, as they may become very dry and hard in fresher surfaces.

Because Trinidad is an island in the tropics, it is likely that chemical weathering is the dominant process by which bedrock is broken down. Chemical weathering is characterized by the loss of mobile elements, that is, elements that are soluble in water such as potassium (K) and sodium (Na). Elements that are less soluble are largely retained as clay minerals in the weathered material. The degree of alteration of a rock can be expressed as the ratio between soluble and insoluble elements. During the process of weathering the concentration of insoluble elements in a rock increases while soluble elements are gradually washed out. This can be expressed using the so-called Chemical Index of Alteration (CIA). The CIA value of a sample is calculated on base of the weight percentages of the respective oxides. Therefore, a sample's CIA is a value between zero and one, with one representing the maximum alteration. The CIA values of clay samples provide insight in their state of alteration, but the CIA values of potsherds are not meaningful because sherds are not a product of weathering. The CIA values have been plotted against each sample's calcium oxide (CaO) concentration.

Trace elements are usually found in accessory minerals. Therefore, grain size can strongly influence a sample's trace element content. For instance, the element zirconium (Zr) is preferentially incorporated in the mineral zircon and the element

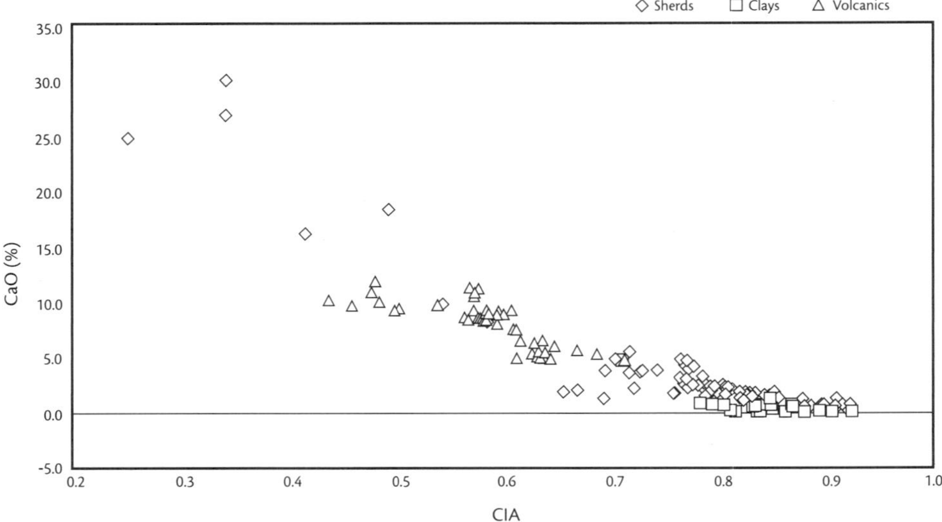

FIGURE 74. The calcium oxide chemical index of alteration (CaO–CIA) values of the potsherds and clay samples from Trinidad compared to those of volcanic rock samples from Saba, St. Eustatius, Martinique, and St. Vincent. $R^2_{sherd} = 0.92$; $R^2_{clay} = 0.69$.

thorium (Th), for example, in monazite. A characteristic of accessory minerals is that they are more resistant to chemical weathering than major rock-forming minerals such as feldspar and amphibole. This leads to the conclusion that clays that have been formed *in situ* as a result of chemical weathering contain a leftover fraction of relatively coarse, resistant crystals. Normally, this fraction is dominated by quartz crystals, enriched by trace elements. Because a clay's trace element content can be strongly influenced by sorting effects, absolute concentrations are unreliable for the comparison of clays. Therefore, trace element ratios instead of concentrations have been used. These were calculated by dividing one element's concentration over another's. Because concentrations of trace elements depend on the same factor (grain size), a clay with a high or low concentration in one trace element is very likely to contain high or low concentrations in other elements as well.

Ratios between trace elements are considered more or less constant in clays from a common source rock. Yttrium (Y) and Zr have proven to be good indicators of provenience (Hofman et al. 2008) and these two trace elements were chosen for analysis for their immobile behavior in chemical weathering. They are both insoluble in water under surface conditions. The Y/Zr ratio is used together with barium (Ba) and lead (Pb). These elements were chosen to compensate for any sorting effects that might have altered the element concentrations. Both Ba and Pb are relatively soluble elements and consequently potentially mobile during chemical weathering. However, this possible weathering effect is affected by the naturally occurring variation between islands. Ba and Pb are also sensitive to changes caused by the firing of pots. Therefore, different production techniques may have played a role if there are geographically specific differences in abundance.

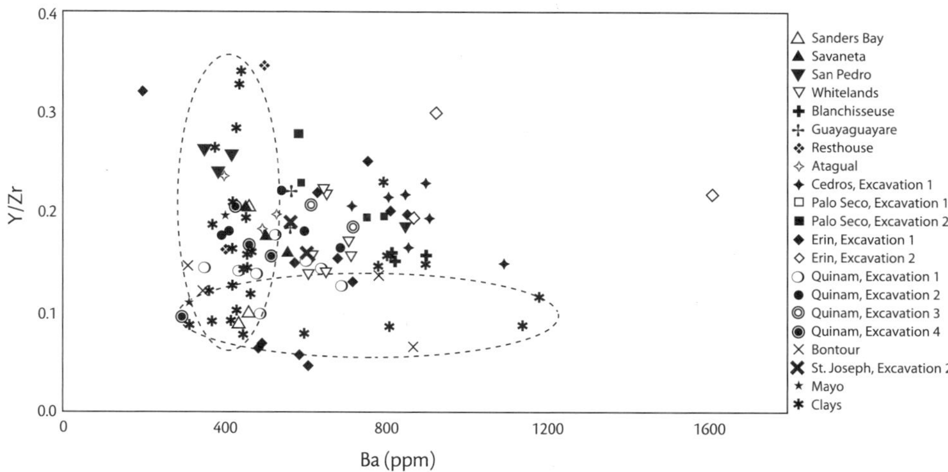

FIGURE 75. The Y/Zr–Ba values of the trace element analysis of all the potsherds and clay samples from Trinidad.

CaO–CIA values were determined for the clays and potsherds from Trinidad compared with those of whole volcanic rock samples from Saba, St. Eustatius, Martinique, and St. Vincent (see Figure 74). (The latter are assumed to be typical of the clay-producing bedrock of the Lesser Antilles.) Most of these volcanics define a tight cluster around a CIA value of 0.60. The clay samples (squares in Figure 74) fall on an approximately straight line extending from a CaO concentration of 0 at a CIA of 0.95. There is a strongly negative correlation between CaO and CIA. The the rock samples fall on a linear trend line through the clay sample population. A trend line based on the potsherd population (diamonds) provides a similar result. The average correlation co-efficient (R^2) of CaO and CIA in clays and potsherds is approximately 0.80 (see Figure 74).

The results of the trace element analysis for the Trinidad clays and potsherds plotted for Y/Zr and Ba values (Figure 75) let us draw several interesting conclusions for the individual sites, as discussed below.

Atagual

The Atagual site of the Palo Seco complex, Saladoid series, yielded two potsherds (UWI) that plot closely together in the whole group and are considered to have a local origin.

Blanchisseuse

Three potsherds (UWI) from the Blanchisseuse site of the Palo Seco complex, Saladoid series all plot closely together on the outside of most of the potsherd and clay sample cluster. They seem to be associated with the whole group and are considered to have a local origin.

Bontour

Four potsherds from Carter's excavation (YPM) at the Bontour site of the Bontour complex, Arauquinoid series, all plot in the main group cluster and are considered to have a local origin.

Cedros

Eight potsherds from the Cedros site, Excavation 1 (YPM) of the Cedros complex, Saladoid series all show association with the main group cluster. They make up the group of potsherds at the high-Ba end. However, all sherds cluster together in one group. They are considered to have a local origin.

Erin

Thirteen potsherds from the Erin site, Excavation 1 (YPM), and three pieces from Excavation 2 (YPM), of this multicomponent site were submitted for analysis. All but two of the potsherds from Excavation 1, most of which can be assigned to the Erin complex of the Barrancoid series, fall within the main cluster of the potsherd and clay samples. The two outliers (518 and 522), both also Erin complex pieces, which plot at the low end of the main cluster, may still be associated with the whole group, but they do not cluster with the other eleven samples from this trench of the Erin site. Two of the three potsherds from Excavation 2, typically belonging to the Palo Seco complex of the Saladoid series, fall outside the main cluster. These pieces (524 and 526) are considered to have a nonlocal origin.

Guayaguayare

Two potsherds from Pit D (UWI) at the multicomponent Guayaguayare site, both of which can be assigned to the protohistoric Mayo complex of the Mayoid series, form a small cluster within the whole group of potsherd and clay samples. These potsherds are considered to have a local origin on the island.

Mayo

Two potsherds of the protohistoric Mayo complex, collected from the surface of Goggin's excavation (YPM) at the Mayo site, were submitted for analysis. They plot in the main group of potsherds and clay samples, suggesting that they are of local origin.

Palo Seco

Four potsherds from Palo Seco Excavation 1 (YPM), and four pieces from Excavation 2 (YPM), of this multicomponent site were submitted for analysis. All the potsherds from Excavation 1, typically belonging to the Palo Seco complex of the Saladoid series, plot closely together and are considered to have a local origin. Note that they form a cluster with the potsherds from Cedros. The potsherds from Excavation 2, which are similarly assignable to the Palo Seco complex, also form a cluster with slightly higher Y/Zr ratios than those from Excavation 2. Nevertheless, they are considered to have a local origin on the island.

Quinam

Eight potsherds from Quinam Excavation 1, five pieces from Excavation 2, two potsherds from Excavation 3, and four potsherds from Excavation 4 (all YPM) were analyzed and plotted. All sherds from Excavation 1, mostly of the Erin complex of the Barrancoid series, cluster closely together in the center of the main potsherd and clay sample group. They are considered to have a local origin. The sherds from Excavation 2, typically assignable to the Palo Seco complex of the Saladoid series, form a cluster and plot next to the eight pieces from Excavation 1, albeit showing slightly higher Y/Zr ratios and Ba values. They are considered to be of local origin. The same applies to the potsherds from Excavation 3, which belong to the Palo Seco complex as well. Both pieces show slightly higher Y/Zr ratios and Ba values than the sherds from Excavation 2. The pieces from Excavation 4, typically belonging to the Bontour complex of the Arauquinoid series, plot within the main cluster and show slightly lower Y/Zr ratios and Ba values than those from Excavation 1. They are considered to be of local origin.

Resthouse

Two potsherds (UWI) from the Resthouse site of the Bontour complex, Arauquinoid series, were submitted for analysis. One of them plots in the center of the main cluster. The other (236) shows a high Y/Zr ratio, but is still associated with the group.

St. Joseph 2

Six potsherds from Excavation 2 (YPM) of the multicomponent St. Joseph site, all of which can be assigned to the Bontour complex of the Arauquinoid series, plot together as a cluster within the main group of potsherd and clay samples and are considered to be of local origin.

Sanders Bay, Chacachacare

All four potsherds (UWI) from the Sanders Bay site of the Palo Seco complex, Saladoid series, seem to be associated with the clays present in Trinidad. They form a cluster in the center of the main group of potsherd and clay samples. It is likely that these potsherds have a local origin.

San Pedro

Three potsherds (UWI) from the San Pedro site of the Bontour complex, Arauquinoid series, plot closely with the main cluster of potsherd and clay samples. They are considered to be of local origin.

Savaneta

Three potsherds (UWI) from the protohistoric Savaneta site of the Mayo complex, Mayoid series, cluster closely together within the main group and can be considered to be of local origin.

Whitelands

All seven potsherds (UWI) from the Whitelands site of the Palo Seco complex, Saladoid series, cluster together with the main group and can be considered to be of local origin.

Conclusions

The XRF analyses of the Trinidad potsherd and clay samples show that by far most of the prehistoric and protohistoric Amerindian potsherds submitted for analysis are of local origin. This applies to all archaeological complexes that can be distinguished. All the more interesting are the few outliers (see Figure 75) that probably represent exotics, that is, vessel fragments from beyond the neighborhood of the site. The two pieces from Erin, Excavation 2, which definitely have a nonlocal origin, plot reasonably close to a clay sample collected in Trinidad's Northern Range and may derive from this area.

The Historic Period Artifacts
Spanish Ceramics and Assorted Tin-glazed Wares

Annette Silver and
Birgit Faber-Morse

The 1953 excavations by Irving Rouse and John and Rita Goggin yielded historic period artifacts recovered from four sites: Bontour, Mayo, St. John, and St. Joseph 2. Six groups of imported pottery can be identified at these sites: Spanish *majolica* ceramics, Spanish earthenware olive jars (*botijas* or *jarras de aceite*), assorted tin-glazed wares, English ceramics, central European ceramics, and Oriental Export porcelain. Other artifacts include pipes, glass, small finds, and architectural materials. (The non-Spanish ceramics and other artifacts are discussed in Appendix C.)

This analysis of the Rouse–Goggin collection is not the final accounting of the historical artifacts recovered from the four sites. A complete treatment awaits incorporation of the data from the Boomert, Anderson-Córdova, and Maharaj/Williams collections now housed in Trinidad at the University of the West Indies. However, the comprehensive study of the St. Joseph 2 ceramic assemblage shows the value of re-analysis of these collections despite the perceived limitations of broad stratigraphic control, small sample size, poor artifact condition, and site disturbance.

The historic period assemblage reflects the presence of the Spanish in the town of St. Joseph (San José de Oruña) and the Mayo mission (Nuestra Señora de Montserrate). The information on the Spanish presence is based primarily on the St. Joseph 2 ceramic assemblage, because four-fifths of the notably few Spanish artifacts in the Rouse-Goggin collections were recovered from this site. The present re-analysis applies dating assignments not available to John Goggin, therefore dates reflect archaeological research after 1970. Re-examination identified sixteen sherds of classified types, fourteen sherds of unclassified decorated *majolica* in the St. Joseph 2 collection, and two unclassified *majolica* sherds in the Mayo collection. Sixty-nine assorted decorated and plain tin-glazed sherds were recovered from the two sites. One Dutch delft sherd was identified among the fragments. Some of the plain and decorated tin-glazed sherds may be English tin-glazed wares, but clear identification as non-Spanish was not possible. The assemblages of Spanish ceramics are in very poor condition and therefore difficult to classify. Our terminology here follows Carruthers (2003), Goggin (1960, 1968), and Lister and Lister (1976, 1982).

The St. Joseph 2 sherds confirm the Spanish presence and direct trade with Spain at the very beginning of the Spanish settlement at St. Joseph in 1592. The

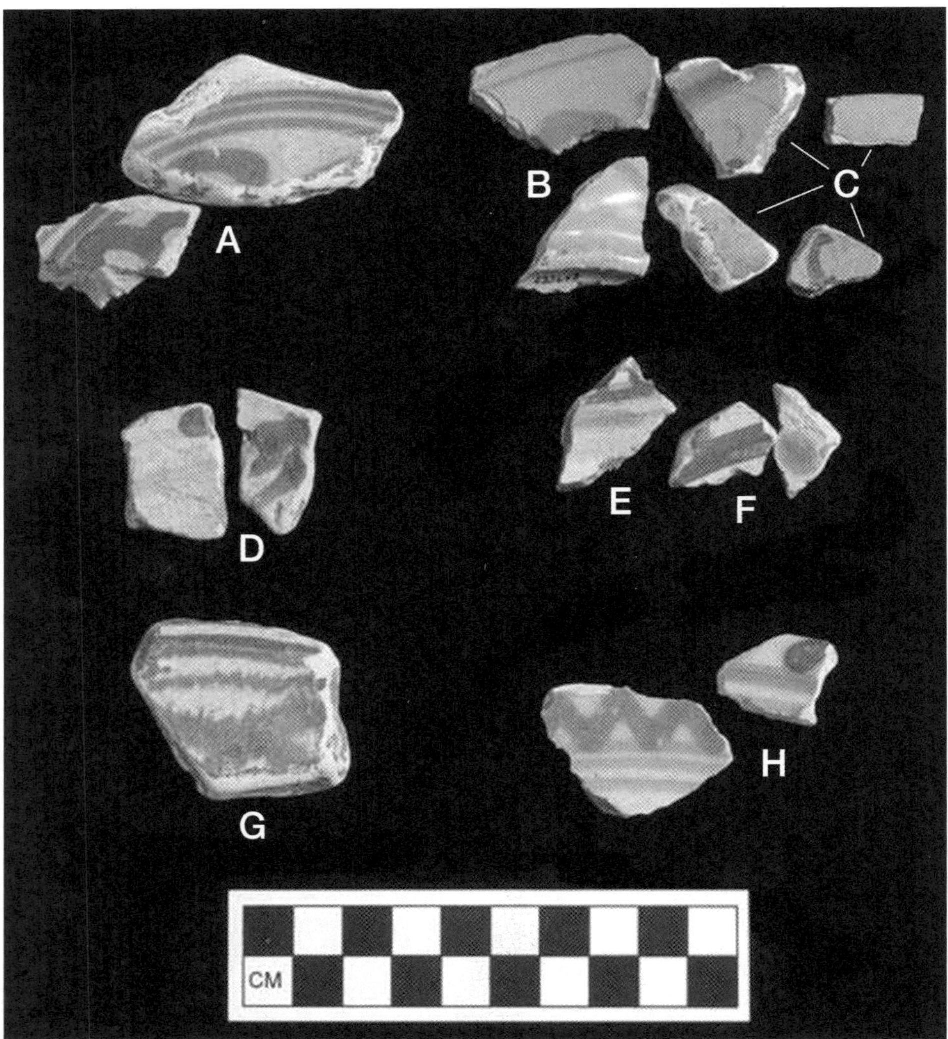

FIGURE 76. Spanish *majolica* and Italian *maiolicas* from the St. Joseph 2 site. *Types:* **A,** Santo Domingo Blue on White. **B,** Sevilla Blue on Blue. **C,** Ligurian Blue on Blue. **D,** San Luis Blue on White. **E,** Aucilla Polychrome. **F,** San Luis Polychrome. **G,** Tallahassee Blue on White. **H,** Puebla Blue on White. *YPM Catalog Nos.:* (A) ANT 255532, 255888; (B) ANT 255648, 255749; (C) ANT 255785, 255786, 255597, 255592; (D) ANT 255792, 255888; (E) ANT 255620; (F) ANT 255577, 255727; (G, H) ANT 255552.

analysis has revealed that the first Spanish settlers in St. Joseph were using *majolica* pottery shipped directly from Andalusia, Spain. Slightly later they were importing a common grade *majolica* made in Mexico City. Shipwreck evidence shows that the early Mexico City San Luis Polychrome and San Luis Blue on White were in circulation throughout the Caribbean until 1622 and 1642, respectively (Marken 1994:233–244). Mexican manufactured *majolica* dominated the finer wares from about the 1640s through the eighteenth century. *Fino* and *entrefino majolica* as well as common *majolica*, made in Puebla, Mexico, and in Mexico City (Puebla Blue on White, Aucilla Polychrome), were in use during the later Spanish occupation.

Only one style of *botijas*, the Middle Style, was identified at St. Joseph. The absence of Late Style *botijas*, introduced around 1775, provides material evidence for the move of the most influential Spanish families from St. Joseph to Port-of-Spain before that time. A few *creole* Spanish and *mestizo* families remained, approximately twenty-two landowning families in all; however, these were "impoverished and isolated" (Mallet 1802; Brereton 1981:20, 84–55; see Chapter 2). According to the material evidence, they no longer used the ubiquitous Spanish olive jar and other imported Spanish wares during the thirty years prior to the English takeover in 1797 (Silver and Faber-Morse 2010).

Majolica/Maiolica

During the Caliphate period in Spain (eleventh to thirteenth centuries), the Moors introduced a distinctive thick tin-glaze technique that became known as *majolica* and *loza blanca*. In Italy the use of a similar thick tin-enamel glaze began in the fourteenth century. This ware was known as *maiolica*. As individual potters moved westward through Europe the use of the Italian technique spread, becoming known variously as *faience* in France and Germany, and as delft, Holland ware, and galleyware in Holland and England. During the sixteenth century in Spain many of the distinct Moorish styles and decorative traits were replaced by Italian Renaissance styles (Lange 2001:11). Three Moorish styles did continue in use after this time. This Moorish–Spanish (Morisco) tradition was carried to Mexico during the sixteenth century (Lister and Lister 1982:83–84; Deagan 1987:54). Santo Domingo Blue on White, a variant type of the Morisco ware, can be identified at the St. Joseph 2 site. Using data from stratigraphically controlled archaeological assemblages, Lister and Lister (1982) categorized the Spanish *majolica* recovered in the New World according to locale of manufacture and the quality of the ware (*fino*, *entrefino*, and *comun*, or common, grades). In our analysis we refer to both Goggin's seminal typologies and to the Listers' categories (Goggin 1960, 1968; Lister and Lister 1976, 1982).

Goggin (1968:41) counted twenty-eight *majolica* fragments in the St. Joseph 2 assemblage, sixteen from Trench 1 and twelve from Trench 2. He identified a group of unclassified *majolica* sherds and seven different *majolica* types: Santo Domingo Blue on White, Ichtucknee Blue on Blue, San Luis Blue on White, San Luis Polychrome, Aucilla Polychrome, Puebla Blue on White, and Tallahassee Blue on White. Re-examination identified sixteen sherds of classified types and fourteen sherds of unclassified decorated *majolica* in the St. Joseph 2 collection, and two unclassified *majolica* sherds in that from Mayo. The Spanish and Mexican *majolica* represented only 1.1% of the St. Joseph 2 ceramic assemblage.

Spanish and Italian Majolica

The *majolica* type Santo Domingo Blue on White is one of a ware group named

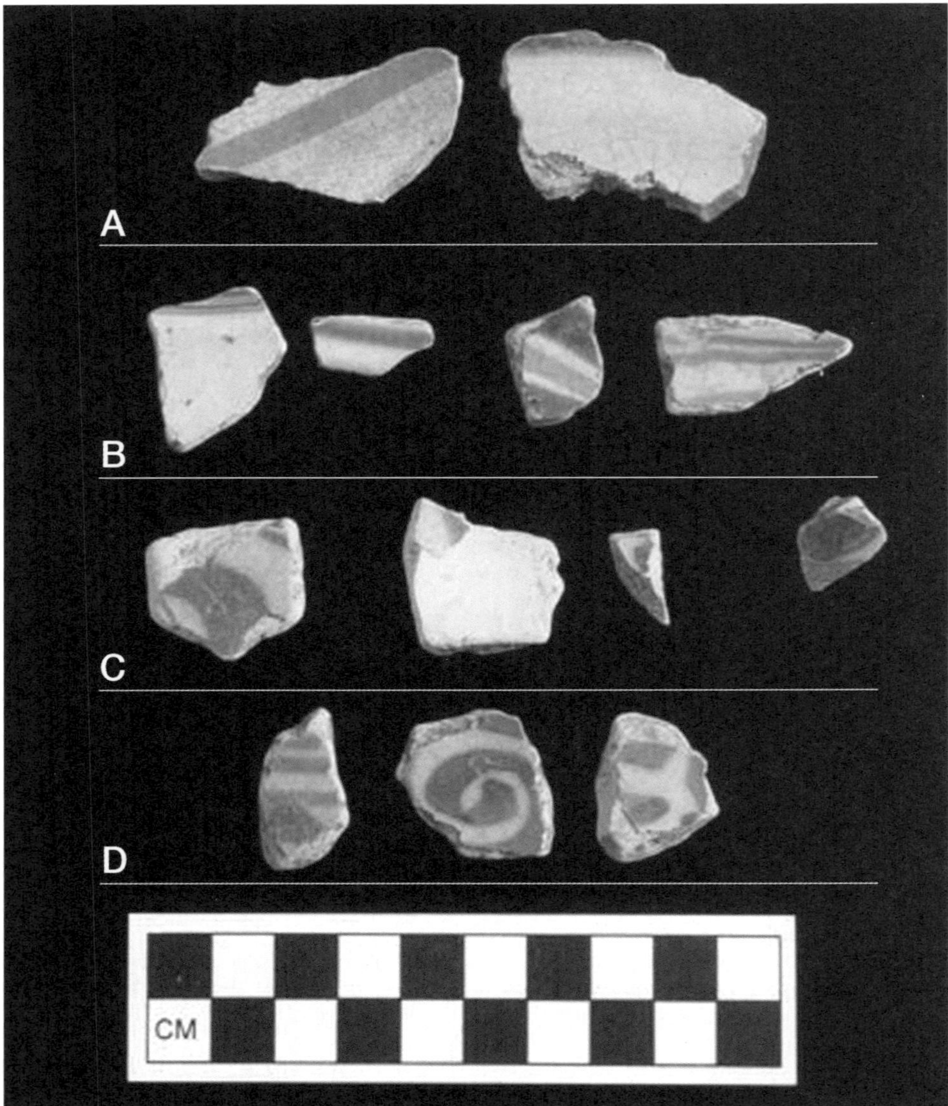

FIGURE 77. Unclassified Spanish *majolica* sherds from the St. Joseph 2 site. *YPM Catalog Nos.:* (Row A) ANT 255581, 255620; (Row B) ANT 255736, 255620, 255734, 255888; (Row C) ANT 255847, 255901 (rim), 255577, 255826; (Row D) ANT 255785, 255649, 255577.

Morisco ware by Lister and Lister (1982) (Figure 76A). This pottery type was manufactured in the Seville area of Andalusia. Santo Domingo Blue on White is the second most abundant common grade ware identified at Spanish colonial sites of the early period (Deagan 1987:70, Figure 76C). Goggin (1968:41) associated this pottery with the "first Spanish occupation" at St. Joseph. Current assessments for New World contexts of this type place its use between 1550 and about 1630–1640 (Deagan 1987:61). Marken (1994:224) notes the recovery of this ware from shipwrecks occurring in 1554 and 1622.

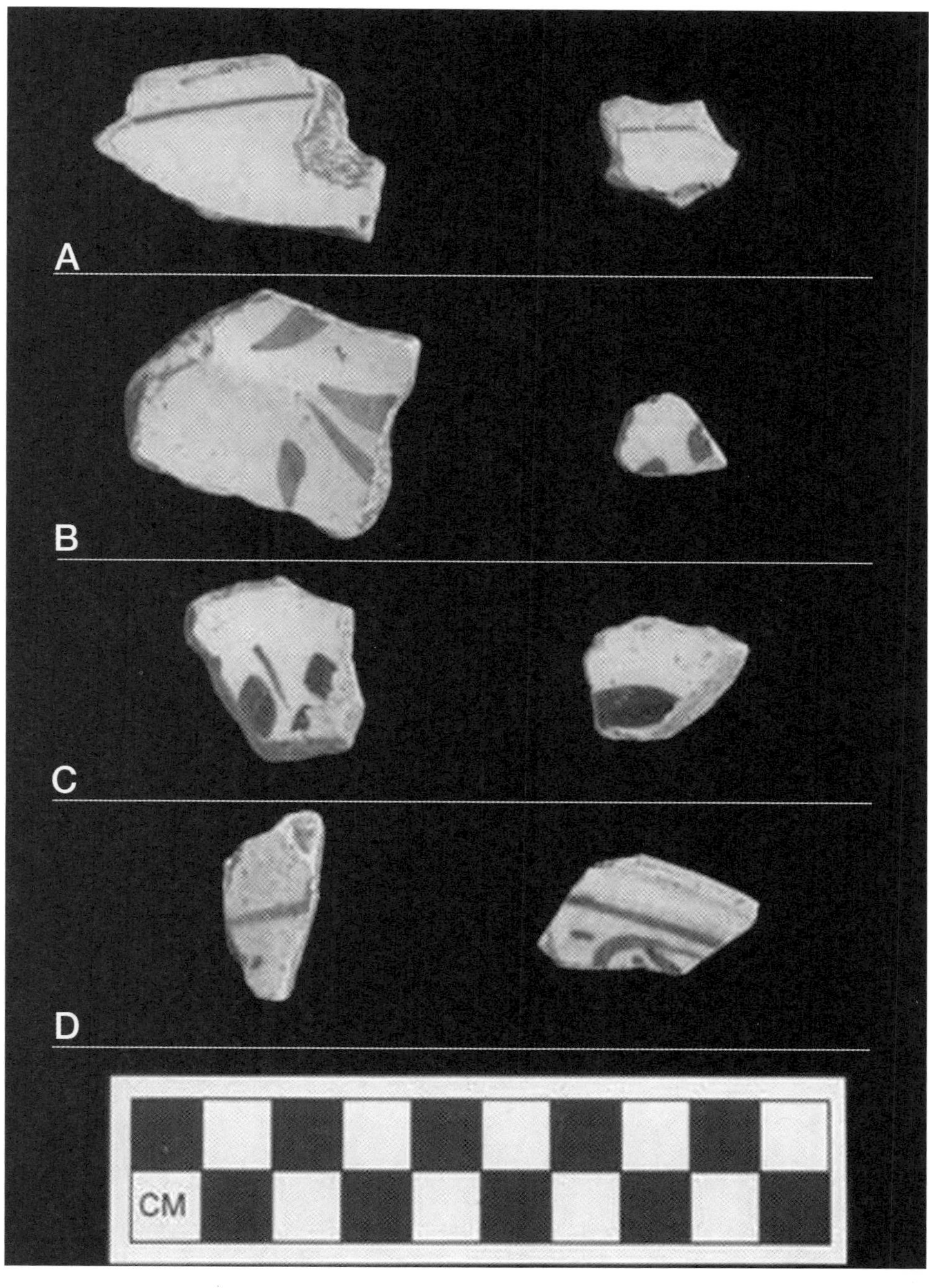

Figure 78. Assorted European tin-glazed ceramics from the St. Joseph 2 site. *Types:* **B, left,** Dutch delftware. *YPM Catalog Nos:* (A) ANT 255749 (left), 255536 (right); (B) ANT 255749 (left), 255552 (right); (C) ANT 255577 (left), 255536 (right); (D) ANT 255642 (left), 255597 (right).

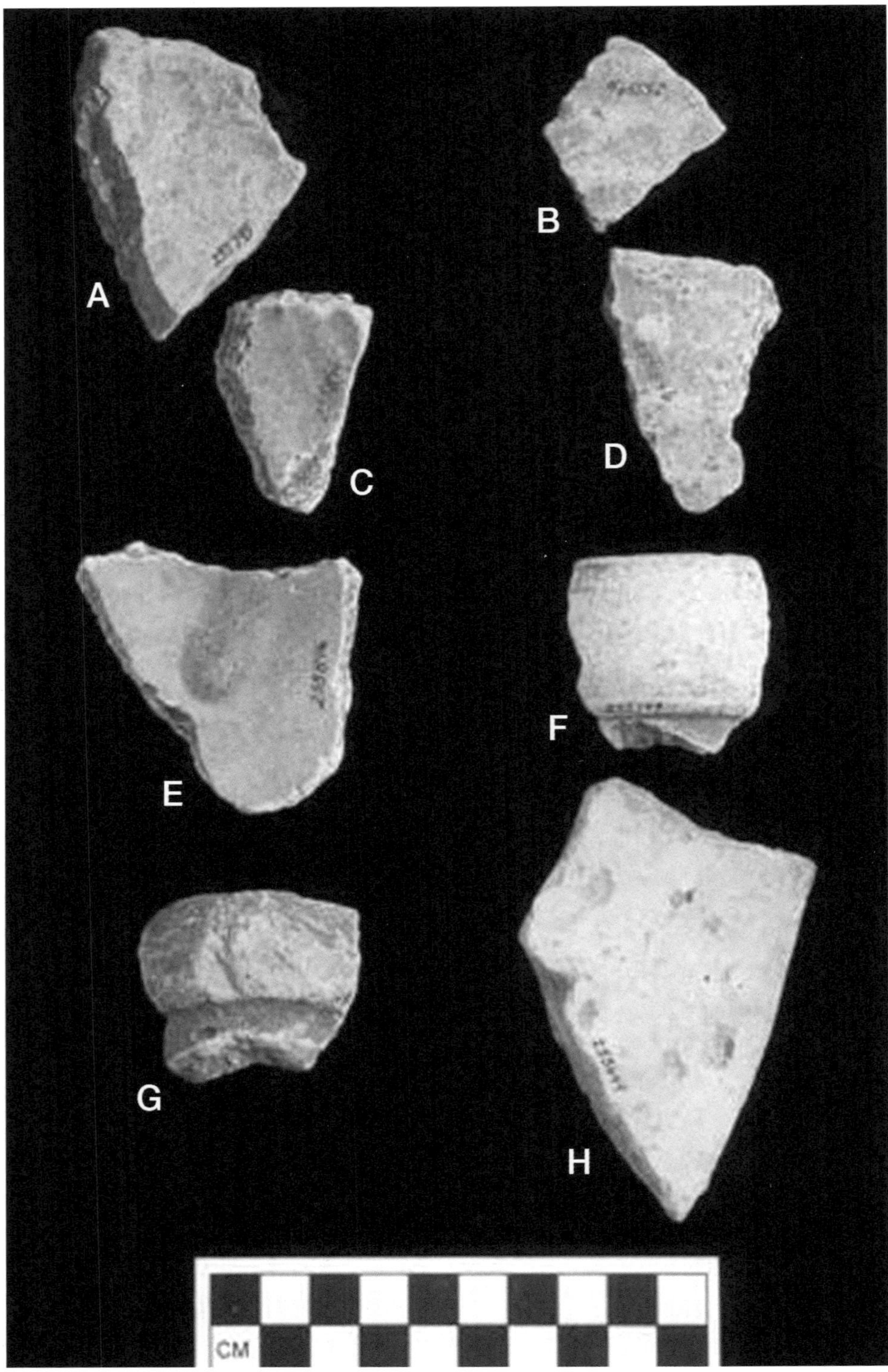

FIGURE 79. Spanish olive jar (*botija*) ceramics from the St. Joseph 2 site. *YPM Catalog Nos.:* (A) ANT 255799; (B–E) ANT 255896; (F, G) ANT 255799 (rims); (H) ANT 255649.

The re-examination identified six sherds as Ichtucknee Blue on Blue. In subsequent analyses of other assemblages, Goggin's original Ichtucknee Blue on Blue type has been resorted into two types, Sevilla Blue on Blue and Ligurian Blue on Blue. According to Deagan, Sevilla Blue on Blue (Figure 76B) was manufactured in the environs of Seville and dates from about 1550 to 1630–1640 in New World contexts. The ware was a very popular export around 1600 (Deagan 1987:62–64). The Ligurian Blue on Blue ware was made in Italy's Liguria province and dates from 1550 to 1600 (Lister and Lister 1982:62, 72; Deagan 1987:70; Figure 76C). Goggin (1968:41) associates his two sherd categories with the "first Spanish occupation" at St. Joseph. The more recent date assessments confirm the Spanish presence and direct trade with Spain at the very beginning of the Spanish settlement of the town in 1592.

Mexican Majolica

The types San Luis Blue on White and San Luis Polychrome were manufactured by European Spanish potters working in the environs of Mexico City. San Luis Blue on White is a *fino* grade ware (Figure 76D). This type is now dated earlier than Goggin's initial assessment and is considered to have been in circulation throughout the Caribbean from 1550 to at least 1650. This end date is based on its recovery from a 1642 shipwreck (Lister and Lister 1982:13–18; Marken 1994:233–234). The common grade ware San Luis Polychrome is a "more careful" treatment of Mexico City Green on Cream (Figure 76F). Shipwreck finds suggest that export of this ware from Mexico City was underway as early as the 1640s (Marken 1994:178, 233–234). It had a sparse presence in Mexico City throughout the second half of the seventeenth century (Lister and Lister 1982:28).

Aucilla Polychrome is another type made in Mexico City (Figure 76E). This common grade pottery was made from 1650 to 1700, with peak popularity in Mexico City from 1680 to 1685 (Lister and Lister 1982:76–77). Puebla Blue on White (Figure 76H) is a fine ware (*fino* and *entrefino* grades) made in Puebla, Mexico. This was a long-lived ceramic tradition, in use between 1675 and 1830 (Lister and Lister 1982:76–77; Deagan 1987:84). Marken (1994:236) finds Puebla Blue on White in shipwrecks from 1724 and 1733. His evidence for trade circulation of Puebla Blue on White and this ware's widespread Caribbean circulation around 1730 suggest its use in St. Joseph in the early 1700s.

Other Types Identified

Tallahassee Blue on White (Figure 76G) is a scarce type recovered from sites in Florida, New Mexico, the Dominican Republic, and Trinidad in associations dated 1635 to 1700, but it is not found in Mexico. However, Goggin proposed that it was made in Mexico, based on physical parameters (Goggin 1968:158–159).

Unclassified Majolica

We counted thirteen *majolica* sherds of Unclassified Blue on White and one unclassified Black on Emerald in the collection of the St. Joseph 2 site (Figure 77). Two *majolica* sherds were recovered from the Mayo site, one sherd of Unclassified Blue on White and one sherd of Unclassified Gray Blue on White.

Assorted Tin-glazed Ceramics

The St. Joseph 2 assemblage contains twenty-five plain tin-glazed sherds, one Dutch delft sherd, one eroded sherd, and thirty-six unclassifiable decorated tin-glazed sherds. The latter may include some English tin-glazed wares (Figure 78). At Mayo, seven un-classifiable tin-glazed sherds were recovered: one plain rim sherd, two plain mouth sherds from a container or bottle, three plain body fragments, and one small decorated (Blue on White) fragment. All were recovered from the upper 20 cm of Trench 3.

Olive Jars

Goggin (1968:41) refers to the presence of "numerous" sherds of olive jars (*jarras de aceite, botijas, botijas peruleras*) of the Middle Style or type at the St. Joseph 2 site. The Middle Style is distinguished by its thick vessel walls (10 to 12 mm) and ring mouth (17 to 26 mm). There may be stamping on the ring mouth. Surfaces may be plain, interiorly glazed, or both interiorly and exteriorly glazed. Handles are absent. Among the three olive jar types identified by Goggin, paste, vessel size, vessel capacity, and finish are all nondiagnostic. The slip, variations in the green glaze tint, and absence of enameling are nondiagnostic traits. These jars would have likely been manufactured in southern Spain, possibly Andalusia, and the type is commonly recovered from Venezuela to Georgia in the United States (Goggin 1960:15–16).

Olive jar sherds are easily visible in the collections. We counted twenty-one sherds from the St. Joseph 2 site and one from Mayo (Figure 79). The St. Joseph *botijas* include two rim sherds and represent, at most, a minimum number of vessels count of three vessels based on paste and rims. This is not the "numerous" subset noted by Goggin (1968:41). These few sherds are comparable to the sparse findings (literally "a few sherds") at the large mission sites in Texas, New Mexico, and Arizona in the United States, as mentioned by Goggin (1960:17). However, the assemblage size is drastically lower than the assemblage of 585 rim sherds recovered at the Santo Domingo Monastery site in Guatemala (Carruthers 2003). There are no marks on the St. Joseph rim sherds, but this observation is not significant because of the small sample size. The buffware bodied sherds at St. Joseph 2 were recovered from all depths of both excavation trenches. In contrast, the salmonware bodied sherds were recovered only from the 10 to 20 cm level of Trench 2.

Botijas were imported from Spain throughout the period of Spanish Indies trade. On the basis of the archaeological date generated by Carruthers and Marken, the date range for Middle Style *botijas* is 1500 to 1773, with very few finds associated with the period between 1500 and 1550. Carruthers pushes back Goggin's 1780 date for the transition from Middle to Late Style to at least 1773, reflecting finds at the Santo Domingo monastery, Guatemala (Marken 1994:52–53; Carruthers 2003:41, 52–53).

THE HISTORIC PERIOD ARTIFACTS
ENGLISH AND OTHER IMPORTED CERAMICS, PIPES, GLASS, SMALL FINDS, AND ARCHITECTURAL REMAINS

Annette Silver

Historic period assemblages reflect English trade or an English presence, or both, at all four of the sites excavated in 1953 by Irving Rouse and John and Rita Goggin (Bontour, Mayo, St. John, and St. Joseph 2). Despite the documented histories, there is scant material evidence earlier than the nineteenth century for European settlement, limited presence, or trade other than that of the Spanish. All such evidence comes from the St. Joseph 2 site. There are fragments of an early seventeenth-century vial, a Rhenish saucer or plate (1650–1750), a German stoneware bottle, a mid-seventeenth-century English pipe, a pipe made in the early half of the eighteenth century and a mid-eighteenth-century English pipe, in addition to a few mid-eighteenth-century ceramic sherds.

The historical materials at Bontour represent a mid-nineteenth-century presence of approximately forty years from the beginning of English rule into the 1870s. The small Bontour historic period assemblage reflects primarily the presence of English merchants in the town of San Fernando and the import of English goods for use by local planters. The date range is indicated by the Shell Edge plates (1809–1813 to 1831–1834), mid-nineteenth-century wine bottles and Dutch pipes, and a celluloid vessel base (post-1868).

The influence of English occupation at the Mayo settlement was minor compared to that at the other three sites, according to minimum number of individuals (MNI) and minimum number of vessel (MNV) counts. The dominant elements in this small assemblage are nails, followed by lesser minimum number of individual of smoking pipes, small finds, and beer and wine bottle fragments. There is an almost total absence of ceramics in comparison to the other sites. Only a few items from Mayo can be assigned temporally. These are the two sherds of creamware tableware (1775–1820) and the mid- to late nineteenth-century horseshoes and pipes. These likely reflect the initial period of the post-1867 resettlement at Mayo.

Ceramics (measured by MNI and MNV) represent 36% of the Bontour assemblage and less than 10% of the Mayo and St. John assemblages. The St. Joseph 2 ceramic assemblage, however, is extensive enough (62%) to support a socioeconomic analysis (Silver 2009). The results show that the most intensive importation of English goods into St. Joseph occurred in the early to mid-nineteenth century.

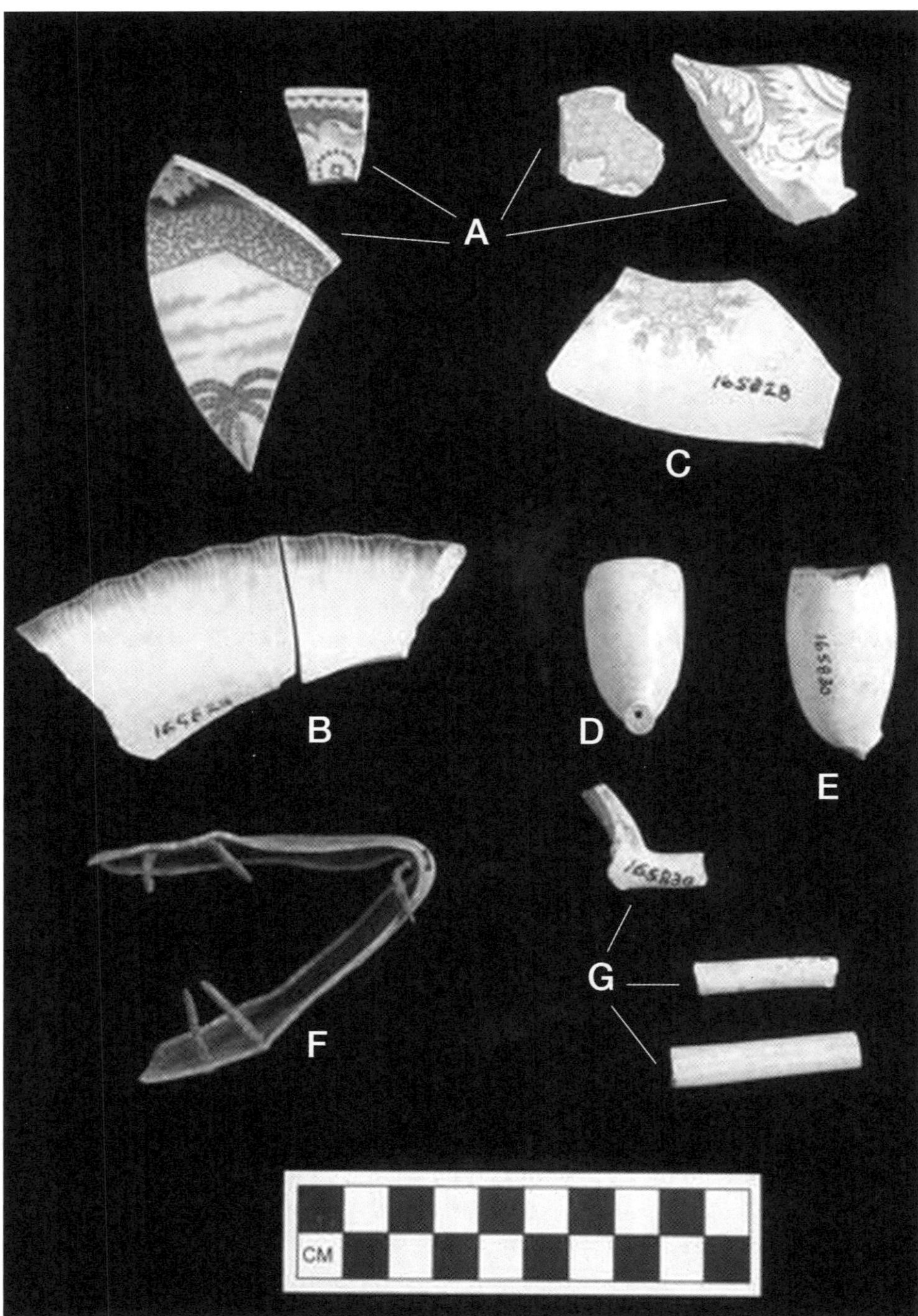

FIGURE 80. Historic period artifacts from the Bontour site. *Types:* **A,** sherds of transfer print plates. **B,** rim of shell edge plate (Miller C). **C,** sherd of transfer print hollowware. **D, E,** pipe bowls. **F,** bail rim. **G,** pipe stems. *YPM Catalog Nos.:* (A) ANT 165992, 165791, two not numbered; (B, C) ANT 165828; (D) ANT 165795; (E) ANT 165830; (F) ANT 165798; (G) ANT 165830 (upper), ANT 165783 (lower two).

The presence of Canton ware at St. Joseph 2 hints at some trade between the English occupants of Trinidad and American merchants during this period. The analysis reveals clear parallels between the St. Joseph 2 English pottery collection and ceramic assemblages recovered from English regimental sites in Canada. The study identifies a regiment economic purchasing pattern and clarifies the specific nature of the nineteenth-century British presence at St. Joseph.

English, German, and Chinese ceramics were identified in the collections. At all four sites the imported English ceramics represent an overwhelming percentage of the historic period artifacts. (Unless otherwise noted, dates used here are the starting and ending dates of manufacture; sources are Miller [1989, 2000], Sussman [1997, 2000], Miller et al. [2000], and Samford [2000].) Rickard's (2006) designations are used for the machine-made ceramics. Note that the literature often lists factory-made slipwares as "dipt," "dipped," "mocha," "banded," "annular," or "industrial slipware." The terminology used in describing the tablewares and teawares follows Sussman (1997) and Miller (2000). The body ware types (creamware, pearlware, and whiteware) post-dating 1820 are considered to be less significant than the decorative types (Sussman 2000:50). Therefore, the decorated creamwares, pearlwares, and whitewares are treated as a separate analytic unit in the following discussion.

The English Ceramics

The English ceramics include a wide range of ware types: utilitarian redwares, graywares, buffwares, and stonewares, refined white earthenwares, and softpaste porcelain. St. Joseph 2, with the largest historic period assemblage, yielded the most extensive range of English ceramic types. At all four sites the ceramics are in very poor condition, totally fragmented with many sherds cracked and burnt (see examples from the Bontour, Mayo, and St. John sites; Figures 80–82.)

Utilitarian Wares

The utilitarian earthenwares (redware, buffware, grayware, yellowware, and tanware) and stonewares make up from 1% to 88% of the English ceramic assemblages. The redware sherds represent plain, lead-glazed, slipped, and lead–manganese glazed vessels. The vessel functions that could be identified are a shallow vessel or platter at Bontour (MNV 1) and crocks, a pitcher, and platters at St. Joseph (MNV 5). No vessel function could be ascertained for the redware body sherds from Mayo and no redwares were recovered at St. John. There are a few plain redware sherds (MNV 1 at Bontour and MNV 3 at St. Joseph) that may or may not have been English imports. Among the St. Joseph sherds, a large plain redware handle suggests use as a storage jar. Smoothed fine plain fragments of a probable flowerpot were also recovered.

Buffwares are rare at the Bontour and St. John sites and absent at Mayo. All are types commonly found at English colonial sites. Two fragments were recovered at Bontour (MNV 2, one painted, one transfer print) and one lead-glazed sherd at

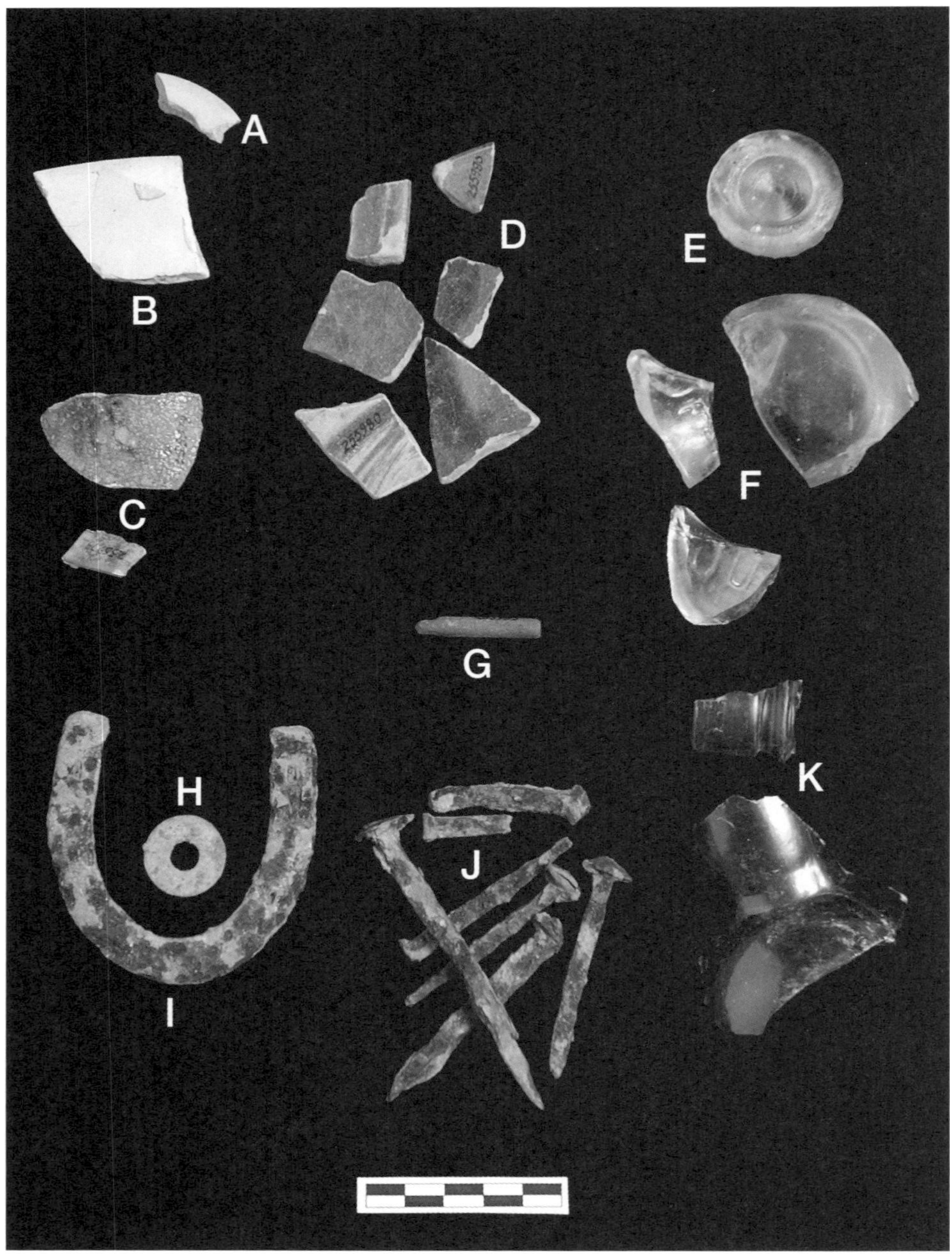

FIGURE 81. Historic period artifacts from the Mayo site. *Types:* **A,** rim of tin-glazed bottle. **B,** rim of tin-glazed pan. **C,** salt-glazed cobalt decorated buffware. **D,** lead and manganese glazed redware. **E,** glass stopper. **F,** washer. **G,** writing lead. **H,** embossed glass container fragments. **I,** horseshoe. **J,** rosehead and L-head wrought nails. **K,** rim, neck, and shoulder of wine bottle. *YPM Catalog Nos.:* (A) ANT 255960; (B) ANT 256014; (C) ANT 255979; (D) ANT 255980; (E) ANT 255944; (F) ANT 255946; (G) ANT 255985; (H) ANT 255944; (I) ANT 255022; (J) ANT 255986; (K) ANT 255984.

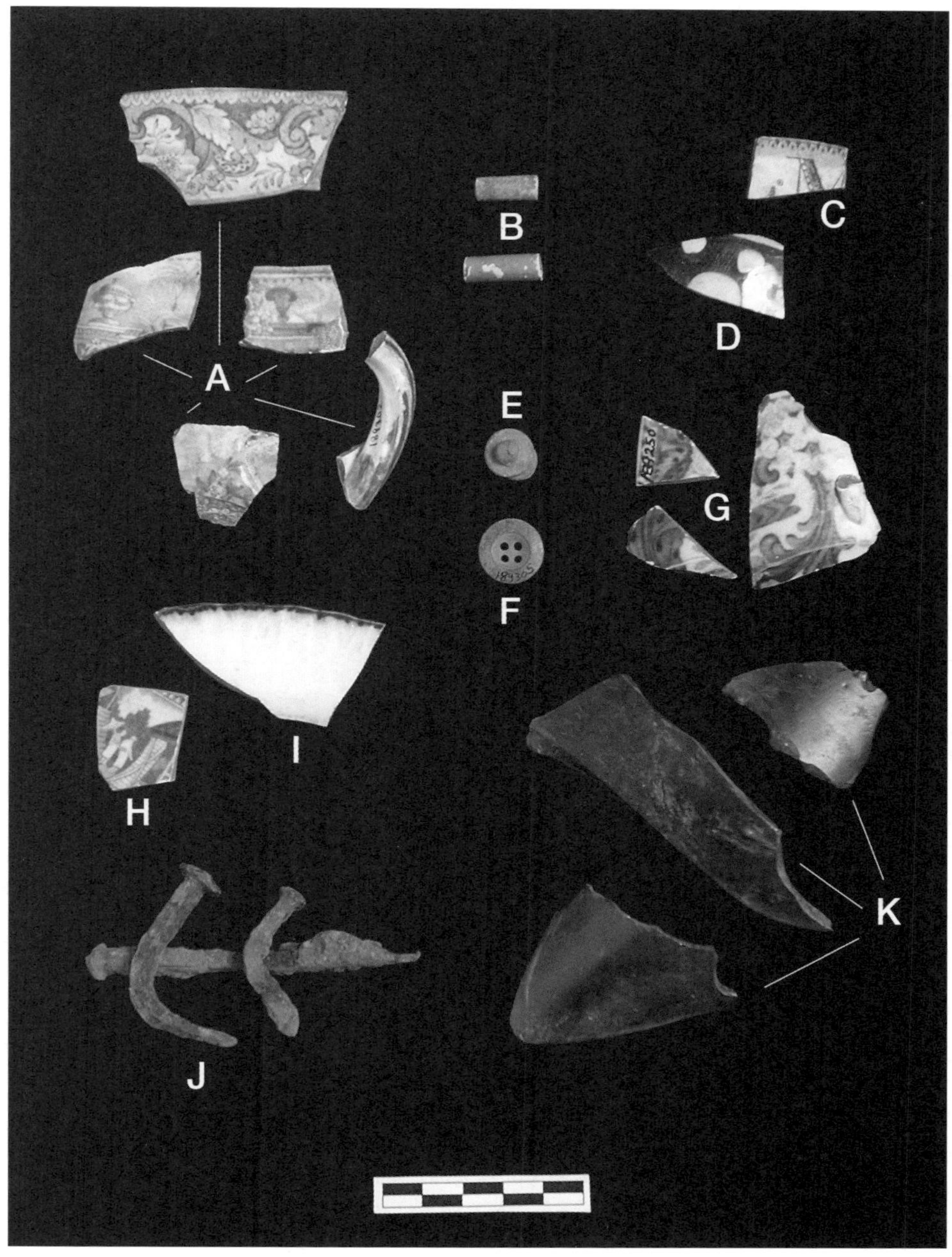

Figure 82. Historic period artifacts from the St. John site. *Types:* **A,** rim, body sherds, and handle of transfer print hollowware. **B,** pipe stem ends. **C,** rim of stippled transfer print plate. **D,** annular and dotted factory-made slipware. **E,** rivet. **F,** brass button. **G,** sherds of transfer print cup. **H,** center of transfer print plate. **I,** rim of feather-edge plate (Miller C). **J,** rose head wrought nails. **K,** wine bottle fragments. *YPM Catalog Nos.:* (A) ANT 189329, 189272, 189302; (B) ANT 189251; (C) ANT 189250; (D) ANT 189329; (E, F) ANT 189305; (G) ANT 189250; (H) ANT 189351; (I) ANT 189302; (J) ANT 189275; (K) ANT 189252.

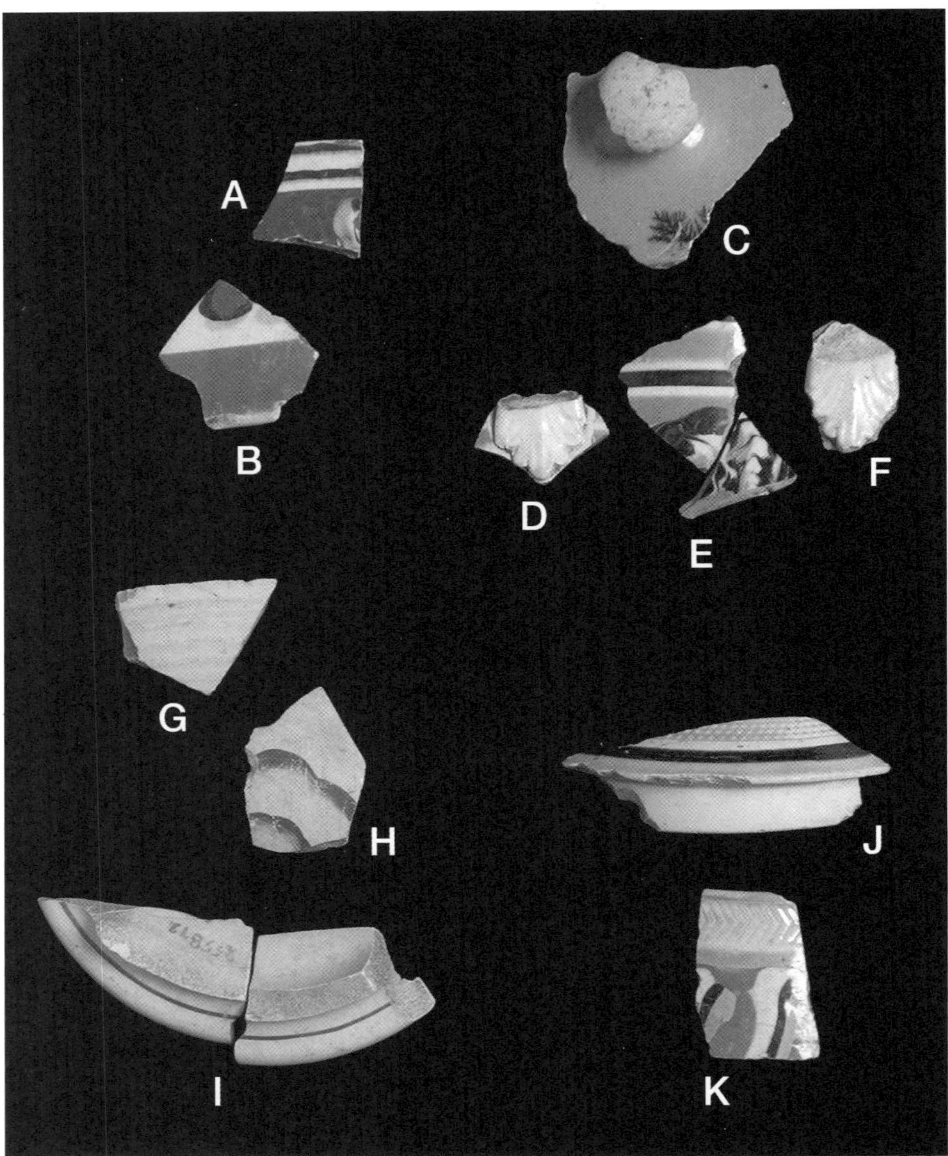

Figure 83. English factory-made slipware and Canary ware from the St. Joseph 2 site. *Types:* **A,** annular and cable decorated rim. **B,** spot decorated rim. **C,** mocha lid. **D–F,** marble-decorated hollowware with handle attachments. **G–I,** Canary ware. **J,** lid. **K,** rim of same vessel (J). *YPM Catalog Nos.:* (A, B, K) ANT 255702; (C) ANT 255560; (D) ANT 255720; (E) ANT 255678, 255624; (F) ANT 255624; (G) ANT 255678; (H) ANT 255627; (I) ANT 255678/255701; (J) ANT 255696.

St. John (MNV 1). At St. Joseph 2, the few buffware body fragments are lead-glazed, lead–manganese glazed, slipped, and salt-glazed (MNV 4). As is the case for redwares, these are types that are commonly found at English colonial sites.

Of note is the presence at St. Joseph 2 of a less common buffware type, represented by fragments of deep yellow-glazed buff earthenware, decorated with molded rilling around the mouth and with intense red scallop designs on the body. The vessel represented (MNV 1) is likely a pitcher or jug (Figure 83G–I). The catalog notes, possibly by Goggin, describe the vessel sherds as "Canary" ware. Canary ware is a term applied to some English yellow-glazed earthenware and in general practice has been used for a range of yellow tones. True English yellow-glazed earthenwares were made on creamware and pearlware bodies and were widely made in England from about 1785 to 1835. They only occasionally appear at archaeological sites in the colonies and the United States (Miller 1974:6, 11). Although the thick deep-yellow glaze and the intense red decoration are comparable to some English pieces (Miller 1974, pl. 8), the St. Joseph 2 vessel is not a classic example of English yellow-glazed earthenware. Its body is a buffware and the decoration is crude compared with the illustrated range of English yellow-glazed earthenware vessels (see Miller 1974).

A yellowware molded rim (MNV 1) and some plain and painted tin-glazed fragments (MNV 4) are other buffware types that were recovered only from the St. Joseph 2 excavation. As noted above, the country of manufacture of these tin-glazed types and delftwares is presently undetermined. A single gray-bodied earthenware handle was recovered at the Bontour site. This vessel may have been made locally.

One stoneware type was identified at Mayo (MNV 1) (Figure 81A). However, at St. Joseph 2, ten varieties of stoneware were identified, representing approximately ten different vessels. There are many functional forms of stoneware vessels, including jugs, storage containers, pots, water jars, and pitchers, to name a few. However, in the St. Joseph assemblage only two crocks and a bottle were clearly identifiable.

Tablewares and Teawares
St. John and St. Joseph 2 are the only sites yielding factory-made slipware sherds (Figures 82D, 83A–F, J, K). There are nine examples from St. John, from the 0 to 20 cm level, and sixty-five from St. Joseph's Trench 1 dispersed throughout all five levels to a 60 cm depth. Because of the multiplicity of slip treatments that can be used on a single vessel, relying on surface treatment alone will overestimate minimum number of vessels totals. Therefore, the minimum numbers of vessels (MNV 3 at St. John and MNV 14 at St. Joseph) are based on different body wares (buffware, pearlware, and creamware) and different rims. The identified vessel functional types are pitchers, a carinated bowl or pitcher, and lidded vessels. A full range of factory-made slipware techniques is represented (annular or slip banding, mocha, marble, separated quill slipping, cable, cat's eye, slip-trailed, feathering, rouletting, and reeding). Manufacture of many of these techniques began in the 1770s and continued into the mid- to late 1800s (the marble, mocha, rouletted, and reeded techniques). Another group (the cable, cat's eye, and feathering techniques) started around 1810 with its highest popularity in the first half of the nineteenth century (Rickard 2006:4–5). Mocha production ceased around 1840 (Miller 2000:91, 101).

Refined whiteware is the dominant tableware category. However, early types are poorly represented and are present only at St. Joseph 2 (MNV 4). There is one white salt-glazed stoneware molded bead-and-reel decorated plate rim, dating from 1720 to 1790 (Miller 2000, table 2). There are a few fragments of Astbury ware (1725–1750) and Whieldon ware (1740–1770) as well as an unclassified greenish-yellow glazed whiteware hollowware (Miller et al. 2000).

Undecorated refined whitewares were recovered at all four sites. (Note that many of the refined whiteware sherds were eroded and not identifiable other than as plain ware.) At Bontour, the plain creamware and plain pearly sherds may be fragments from transfer print vessels at the site. Similarly, the plain creamware and plain pearlware from St. John (base and body sherds) are likely fragments associated with the decorated vessels. At Mayo, however, the plain sherds (MNV 2) are clearly plain creamware fragments as there are no other tableware ceramics. At St. Joseph 2, there is a small cluster of four pearlware plate rims and fragments of two pearlware hollowware vessels (1775 to 1840), molded plain plate rims (MNV 2) and molded plain platter or plate rims (MNV 2). It is likely that the molded creamware plate or saucer and the plain creamware plates were used in the early nineteenth century, in view of their association with a plain creamware carinated hollowware vessel that post-dates 1810.

There are four major categories of decorated wares: shell edge wares, underglazed lined wares, painted wares, and printed wares. Blue and green shell edge plate sherds were identified at Bontour (MNV 3), St. John (MNV 1), and St. Joseph 2 (MNV 10). The latter site also yielded fragments of a charger or platter and a saucer. The shell edge wares provide relatively tight dates of 1809 to 1831 and 1813 to 1834 for Bontour (Miller C and D types) and 1813 to 1834 for St. John (Miller C). At St. Joseph 2, the shell edge ware presence reveals three clusters of English imports and a longer interest in English goods. The three styles Miller, Miller C, and Miller D have beginning manufacture dates of 1802, 1809, and 1813, respectively, and ending manufacture dates in the 1830s (specifically, 1831, 1832, and 1834, respectively). Two recovered types were manufactured later, having date ranges of 1841 to 1857 and 1874 to 1879 (Miller 1989: Miller C, D, E, G, H, blank).

Underglazed lined wares are represented by creamware and pearlware cup fragments (MNV 4), that is, teawares, all from the St. Joseph 2 site. In general, lined wares were made between 1814 and 1833 (Miller 2000:82).

Another group of teawares (cups, an unidentifiable hollowware rim, a lid rim, and handles, MNV 5) are represented by the painted wares recovered from the St. Joseph 2 site. Five stylized floral polychrome patterns (post-1820) are identified in this group. The St. John assemblage yielded a single painted sherd, possibly a remnant of a plate or saucer with a blue painted floral design (MNV 1).

The printed wares (often referred to in the literature as transfer prints) include both tableware (plates and tureens) and teas. A printed creamware hollowware vessel and printed buffware and pearlware sherds were recovered from Bontour (Figure 80A, C). These represent one plate or saucer, a plate or shallow bowl from a Staffordshire pottery, a second hollowware vessel, and two unidentifiable vessels (MNV 6). The printed ware fragments at St. John are of plates, a tureen handle, one flanged cup with handle, an unidentifiable hollowware piece, and one unidentifiable vessel (MNV 6; Figure 82A, C, H). At least two of the patterns seem to be of Staffordshire origin. The Chinoiserie-style patterns, the black stipple transfer printing, and the cup shape provide manufacture date ranges of 1797 to 1814, 1807 to 1830, and 1810 to 1840, respectively (Miller 2000). At the St. Joseph 2 site, the printed ware represents 69% of the decorated refined whitewares and 78% of all refined whitewares. Although the vessel numbers are small, they show great diversity: thirteen separate print-decorated tableware and tea sets are indicated by colors and patterns. The printed wares fall into three groups according to the beginning date of manufacture: Group I, with beginning dates spanning 1775 to 1795 and ending dates spanning 1814 to 1830; Group II, with beginning dates spanning 1807 to 1828 and ending dates spanning 1825 to 1840; and Group III, with an end date of 1870. For a discussion of the social, economic, and historical implications, see Silver (2009).

Porcelains
One painted and molded softpaste porcelain handle sherd was recovered from the Bontour site. This is likely from a teaware set. Twelve sherds of English softpaste porcelain and (probable) English porcelain were recovered from the St. Joseph 2 site. The St. Joseph softpaste ware (MNV 3) includes a bowl or cup and plate or saucer. The porcelain vessels (MNV 2) are polychrome overglazed and gold-tinted overglazed and molded. The latter was a bowl or cup vessel dating from 1790 to 1840. The softpaste wares date from 1745 to 1795.

German, Oriental, and Unprovenienced Delft Imports

A minimum of three vessels manufactured in Germany and China were identified, all recovered from St. Joseph 2. There is evidence for a Rhenish saucer or plate, dated 1659 to 1750, and a German stoneware bottle or flask (Figure 86H); there is a single rim of Oriental Export Canton ware, dated 1800 to 1830. The Canton vessel did not come in with the English trade, because the British East India Company stopped importing Chinese wares in 1791. Its presence at St. Joseph indicates some trade with American merchants, because Americans took over the Chinese trade after the British withdrawal from that market (Miller 2000, table 4). In addition, delft sherds of unclear origin, described already, were recovered.

Pipes

White clay (kaolin) smoking pipes were recovered from all four sites. There are mid-nineteenth-century pipes of Dutch manufacture at Bontour and mid- to late nineteenth-century pipes of Scotch manufacture at Mayo and St. John. The pipes recovered from St. Joseph 2 are of English and Scottish manufacture. They reflect three distinct periods of activity at the town of St. Joseph: the mid-seventeenth

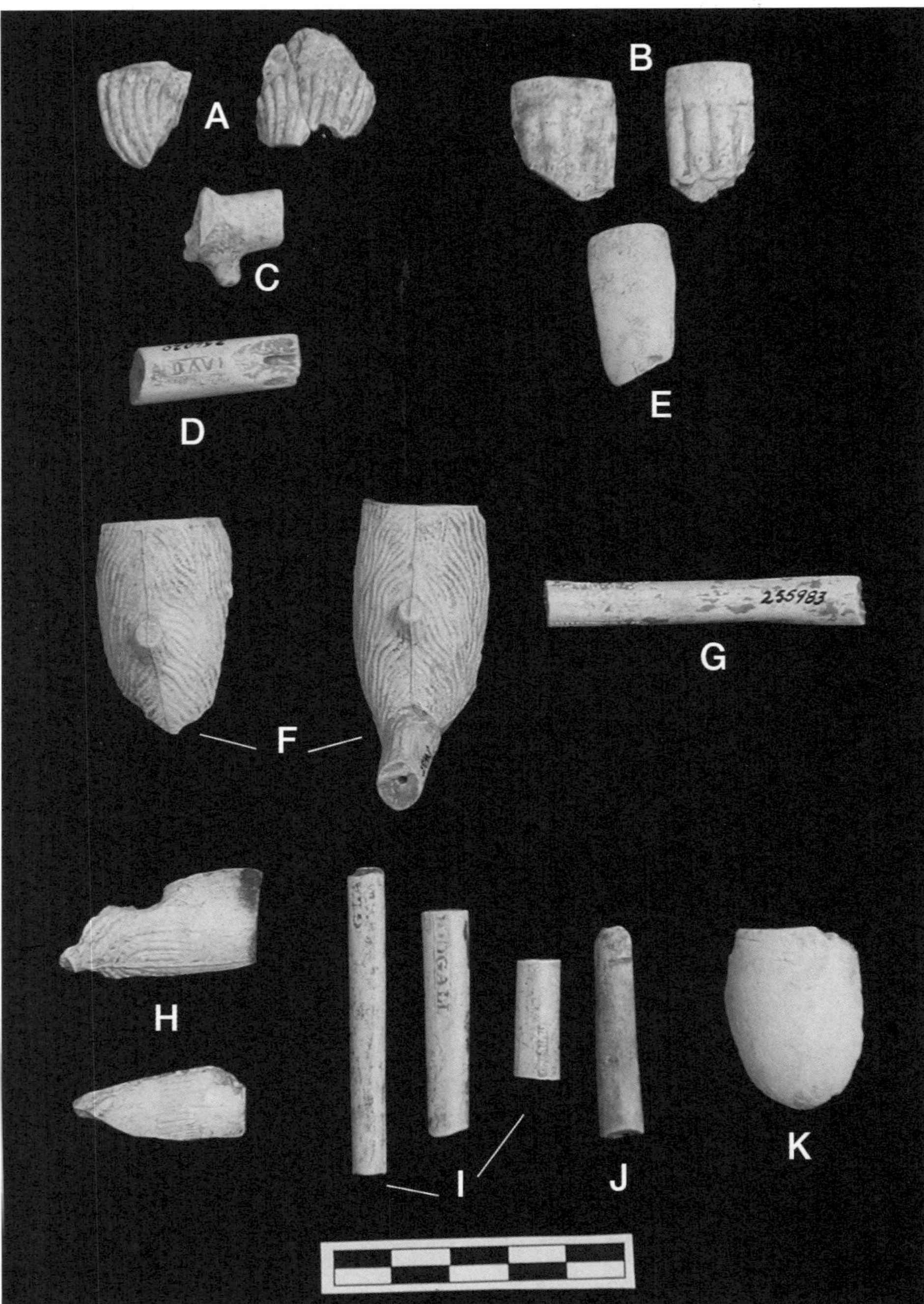

FIGURE 84. Clay (kaolin) pipes from the Mayo (A–G, J) and St. John sites (H, I, K). *Types:* **A,** embossed pipe bowls (MNV 2). **B,** pipe bowl fragments (MNV 1). **C,** pipe spurred heel fragment. **D,** pipe stem "DAVIS//GLASGOW." **E,** plain pipe bowl. **F,** embossed (tree bark) pipe bowl fragments. **G,** pipe stem "Burns/Pipe." **H,** pipe bowl fragments (MNV 2). **I,** three pipe stem fragments. **J,** pipe stem end with bite marks. **K,** plain pipe bowl fragment. *YPM Catalog Nos.:* (A, B) ANT 255019; (C, D) ANT 255020; (E) ANT 256008; (F) ANT 255968, 255967; (G) ANT 255983; (H) ANT 189251 (MNV 2); (I) ANT 189352, 189251, 189304; (J) ANT 256008; (K) ANT 189304.

century (1650 to 1680) and the early eighteenth century (1700 to 1740) with both sets of goods likely brought in as the result of illegal trade with English ships (see comment in Chapter 2), and the mid- to late nineteenth century, associated with English settlement.

Five pipe fragments (MNV 3) were recovered from the Bontour site (Figure 80D, E, G). The bowl shapes, stem attachments, and heel stamps identify these as Dutch elbow angle pipes. The two bowl rims are decorated with fine rouletting. The bases are heelless, spurless, and stamped. The heel stamp on one pipe bowl is a worn unidentifiable mark in a circle under a crown, Dutch, but unidentifiable as to maker (Figure 80D). The heel stamp of the second pipe bowl (Figure 80E), a crowned "16" within a circle, is the stamp of Van der Want, manufactured from 1842 to 1869 plus (McCashion 1979, pl. 16, 37; Azizi et al. 1996:126).

The Mayo site yielded thirteen pipe fragments (MNV 5). Two pipes are decorated with ribbing. On one pipe the bowl is decorated with thin raised ribbing following the contour of the pipe bowl (Figure 84A). The second pipe is decorated with wide vertical ribbing encircling the bowl (Figure 84B). These ribbing decorations are found on pipes from the late nineteenth century and after the 1850s, respectively (Bradley 2000, fig. 17). Two bowls are decorated with a raised "tree bark" design (Figure 84F). There are two pipe stems, one stamped with "BURNS// ...Y PIPE" and the other with "DAVI—//GLASGOW." The first (Figure 84G) was made by Duncan McDougall and Company of Glasgow from 1846 to 1891. The second pipe (Figure 84D) was made by Thomas Davidson and Company, also of Glasgow, from 1862 to 1911 (Azizi et al. 1996:122; Bradley 2000:117).

At the St. John site, forty-one pipe fragments were recovered (MNV 5). Bowls of two pipes are decorated with raised ribbing and date to the nineteenth century (Figure 84H). A third pipe bowl fragment is decorated with barely visible shallow ribbing. It is comparable to pipes recovered from mid- and late nineteenth-century military sites in Canada (Bradley 2000, fig. 17). A fourth pipe is represented by plain bowl fragments (Figure 84K). Three pipe stems bear stamped marks of the Duncan McDougall and Company of Glasgow (Figure 84I). These were made in the period 1846 to 1891 (Bradley 2000:117).

Sixty-four pipe fragments were recovered from the St. Joseph 2 site (MNV 5) (Figure 85A–C, F–H). Two of the pipe bowls are clearly English. A third pipe bowl, roulette decorated, was initially listed as "probable Dutch." However, this bowl fits both the Noël Hume and Mallios illustrations of an early English pipe and is counted here as an English import (Noël Hume 1985, fig. 97:11–12; Mallios 2005, fig. 4:25). The range of manufacture dates indicates three periods of import and use: mid-seventeenth century, early to mid-eighteenth century, and the nineteenth century. Two stamped pipestems, "White//Glasgow" (Figure 85F) and "WHITE??... GO..." identify the nineteenth-century pipe maker William White of Glasgow, Scotland (Oswald 1975; Bradley 2000:117).

The presence of pipes manufactured in Scotland in association with English imports and during the English occupation of Trinidad is not unexpected. Accord-

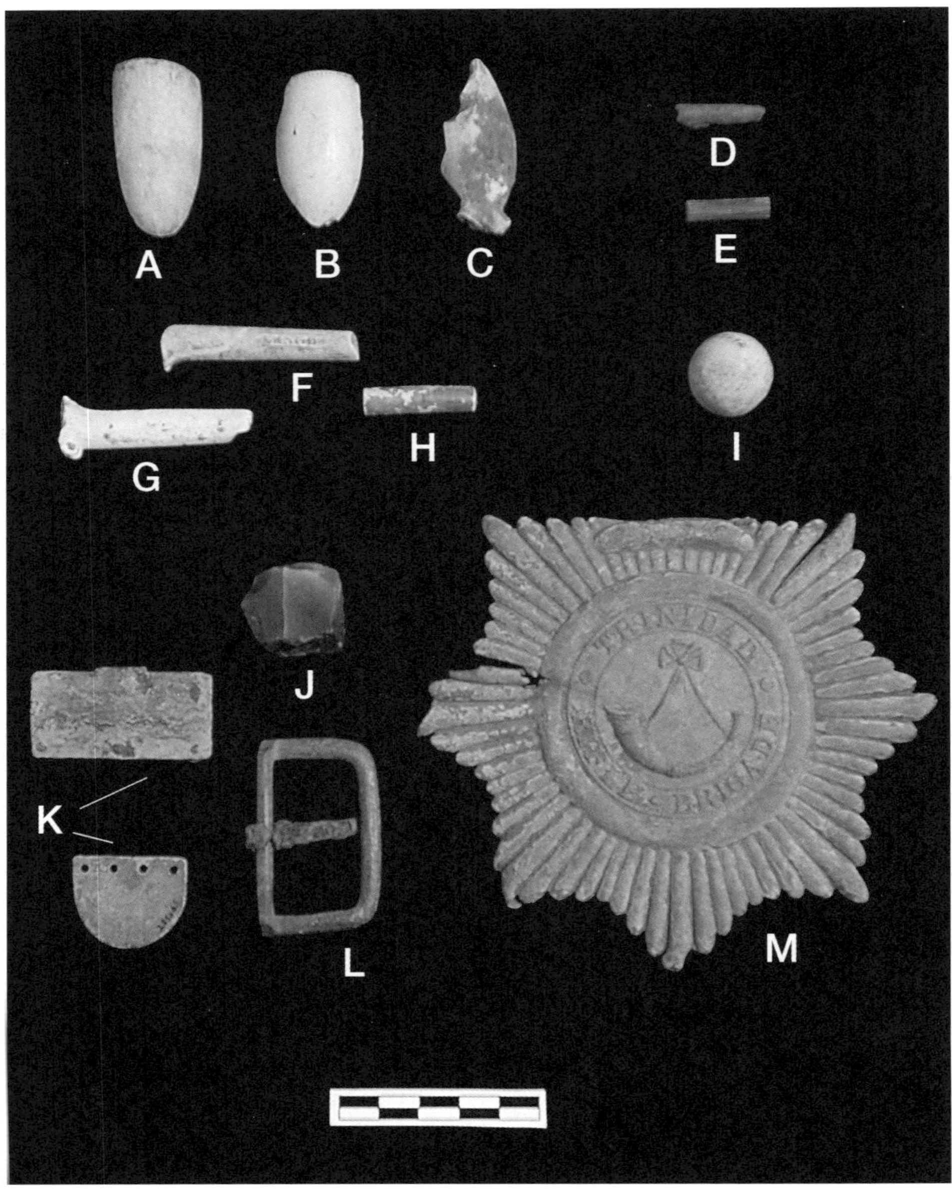

Figure 85. Clay (kaolin) pipes and small finds from the St. Joseph 2 site. *Types:* **A–C,** pipe bowls. **D,** writing lead. **E,** blue glazed bead. **F,** pipe stem stamped "WHITE//GLASGOW." **G,** spurred pipe stem. **H,** varnished stem end. **I,** clay marble. **J,** gun flint. **K,** copper clasp and fastener. **L,** harness buckle. **M,** shield "Trinidad Rifle Brigade." *YPM Catalog Nos.:* (A) ANT 255565; (B) ANT 255613; (C) ANT 255592; (D) ANT 255664; (E) ANT 255663; (F) ANT 255556; (G) ANT 255730; (H) ANT 255745; (I) ANT 255572; (J) ANT 255716; (K) ANT 255715; (L) ANT 255841; (M) ANT 255689.

ing to Bradley (2000:117), Scottish pipe manufacturers "monopolized pipe exports during the 19th century."

Glass, Small Finds, and Architectural Materials

The alcoholic beverage bottle is the only glassware present at all four sites. It is also the most common glass artifact. Of the four sites, St. John and St. Joseph 2 had several glass functional types represented.

Small finds make up a moderate portion of the four assemblages. Similarly, architectural materials represent a minor to moderate part of the assemblages, with the exception of those recovered at St. John. Here, architectural materials represent 70% of the assemblage by MNI. The St. Joseph 2 assemblage as the largest of the four, unexpectedly, does not hold many architectural artifacts. Wrought iron nails are the only architectural materials common to all four sites.

The St. John and St. Joseph 2 assemblages are the largest of the four historic period collections. They are similar in the MNV counts of glass and pipes, the ranges of architectural functional categories (six and four, respectively), and of glass functional types (six and seven, respectively), all possibly a function of large artifact numbers. Nonetheless, among the four sites, the Mayo and St. Joseph 2 sites yielded the widest range of functional categories. This is notable because of the big difference in assemblage size. The MNI and MNV total at Mayo is one-fifth that at St. Joseph.

Glass

Glass vessels are a minor part of the assemblages. At all four sites bottles for alcoholic beverages represent the most common item by minimum number of vessels and at three sites (Bontour, Mayo, and St. John) by fragment counts. Alcoholic beverage bottles are the only functional glass type recovered from all four sites (Figures 81, 82, 86).

There is no pattern to the distribution of the other glassware types other than that the sites with the largest assemblages (St. John and St. Joseph 2) also had a few more functional types represented. Pharmaceutical bottles and vials were recovered only from the St. John and St. Joseph 2 sites (Figure 86A, B). Fragments of container glassware were recovered (a mid-sized container, as indicated by an "Eno's type stopper," and a jar at the Mayo and St. John sites, respectively) (Figure 81E). Glass tableware (stemmed drinking and serving ware) was identified at two sites, Bontour and St. Joseph 2 (Figure 86C–E). Chimney glass fragments were encountered at St. John. Architectural glass was recovered from Bontour, St. John, and St. Joseph 2 (see below). Surprisingly, considering the heavy and durable heft of wine or liquor bases and my experience at other sites, with the exception of the case bottle base there are no wine, beer, or liquor bases recovered from these Trinidad sites. Another anomaly among the glass assemblages is the set of four pieces of utilized and knapped wine bottle glass found at the St. Joseph site. Simi-

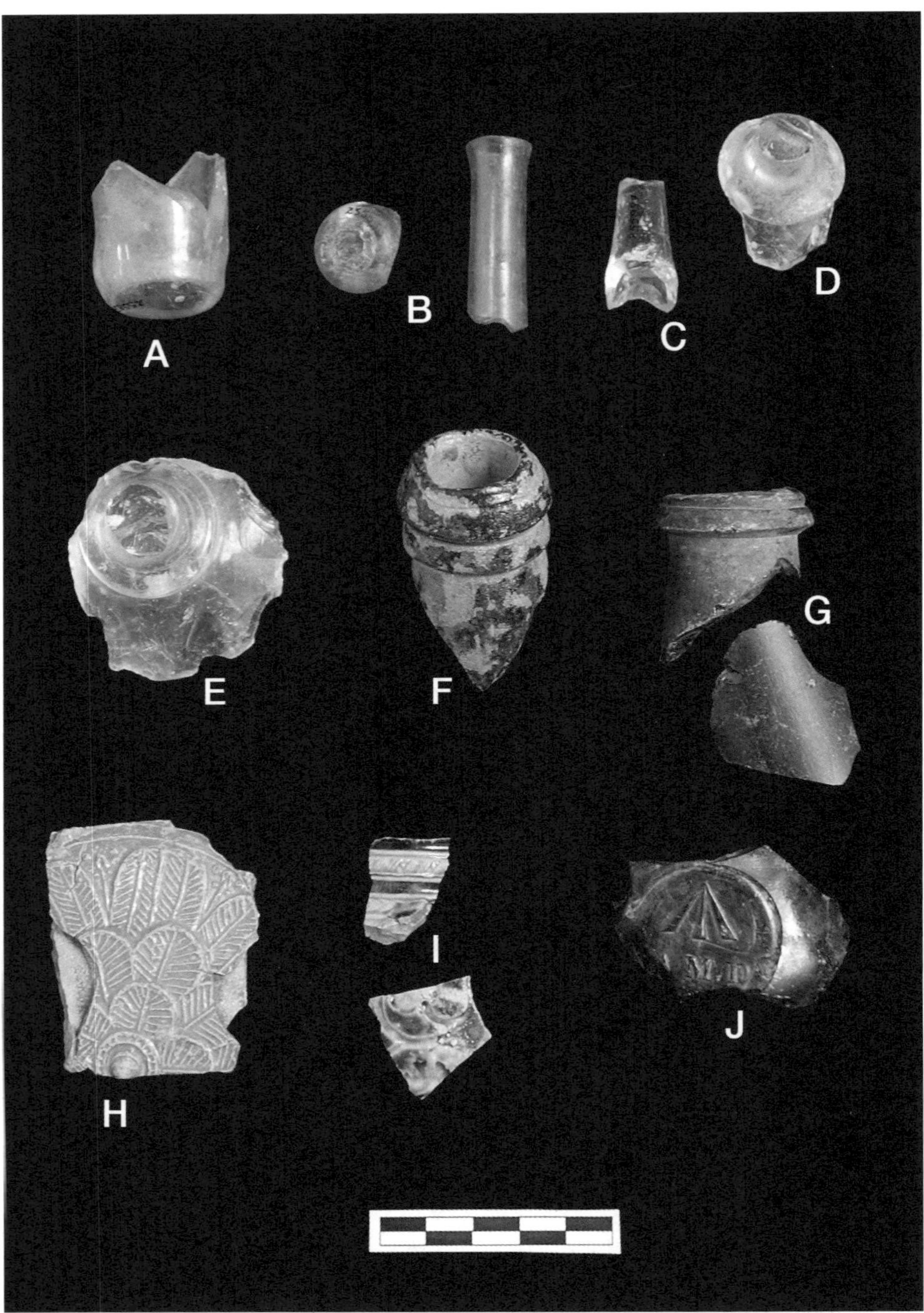

FIGURE 86. Glass and stoneware from the St. Joseph 2 site. *Types:* **A,** pharmaceutical bottle base. **B,** pharmaceutical bottle base and neck. **C,** stemware fragment. **D,** stemware knopf. **E,** blown glass stemware base. **F,** wine bottle finish. **G,** wine bottle finish and shoulder. **H,** stoneware decorative element. **I,** Rhenish rim and body sherds. **J,** bottle shoulder with molded shield. *YPM Catalog Nos.:* (A) not numbered; (B) ANT 255550, 255897; (C) ANT 255897; (D) ANT 255593; (E) ANT 255593; (F) ANT 255573; (G) ANT 255556; (H) ANT 255621; (I) ANT 255750, 255537; (J) ANT 255550.

larly, utilized glass tools have been recovered from early nineteenth-century African American slave household contexts in the United States (Wilkie 2000). Most of the reported tools were made of wine bottle glass, as are the St. Joseph pieces. Wilkie documented common use of broken glass as both razors for personal use and as scrapers to smooth wooden handles. However, she notes that such use of re-utilized sherds may prove to be a "rural" phenomenon, one not necessarily restricted to African American contexts.

One glass item in the Bontour assemblage, a wine bottle finish, provides a date of 1850 to 1859 (Jones 1986:43, ch. 5, table 4). Fragments of wine bottle finishes (MNV 3) from the Mayo site provide no firm dates (Figure 81K). One finish is illustrated among the later bottles of a group dating from 1735 to 1850 (Jones 1986, fig. 49). However, Jones has no dated examples of the illustrated bottle. Two small finish fragments have flat-topped straight lips and may also be of early nineteenth-century manufacture. There is a stoppered container bottle that may have held a cough compound, but no tight date range is ascertained at present (Figure 81E, H). A very tentative date range (1793 to 1836) is suggested for the St. John glass assemblage. This is based on a single rounded "wine" bottle lip (Jones 1986, appendix A).

For the St. Joseph 2 glass assemblage one wine bottle finish provides a date of 1766 (Jones 1986, fig. 25); another finish dates from 1800 to 1837 (Jones 1986, fig. 51). The suggested date for the stemware fragments (a bladed knop, three-part ovoid bowl, and a domed and folded base with pontil; Figure 86C–E) is post mid- to late 1800s (Jones 1986, fig. 76). The suggested date for the pharmaceutical bottle is post-1780 (Noël Hume 1985, fig. 17: 1780 illustration). The applied glass seal with molded "—MD" (Figure 86J) would most commonly appear on bottles pre-dating 1840 (Jones and Sullivan 1989:16–17). The bottle fragments providing the two earliest dates were both recovered from Trench 2 of the site.

Small Finds

This category includes miscellaneous metal kitchenware, arms and ammunition, clothing, personal items, furnishings, and other small items that can relate to a range of activities. The widest ranges of small finds were recovered from Mayo and St. Joseph 2. Nonetheless, small finds are only a minor part of each assemblage.

Fragments of an iron container or pail and a section of an iron circle, bracelet size, were found at Bontour (Figure 80F). Celluloid container fragments, one writing lead, an iron washer, and two horseshoes were recovered from Mayo (Figure 81F, G, I). An approximate date for the horseshoes is the mid-nineteenth century (Noël Hume 1985, fig. 74:7); the celluloid is post-1868 (Miller et al. 2000). Clothing is represented at St. John by a one-hole bone button and a four-hole brass button (Figure 82F), both types commonly in use from 1837 to 1865 (Noël Hume 1985, fig. 23, types 15 and 32). Other small finds recovered from this site were one brass brad, a section of an iron kettle or pail bail, and an iron lock cover. A beveled lead square and an iron disc are identifiable among some iron and lead scraps.

Seven functional categories of small finds were recovered from the St. Joseph 2 site: toys, clothing, arms, writing-related items, furnishing, hardware, and livestock (Figure 85D, E, I–M). Dates from the late eighteenth century to the late nineteenth century were provided by a pony shoe (late 1700s to 1800s), a porcelain doll's arm (1850 to 1880), and the "Trinidad Rifle Brigade" insignia (late nineteenth century).

Architectural Materials
Overall, architectural materials represent a very minor part of the historic period assemblages. Wrought iron nails are the only architectural materials common to all four sites. The rose head is the predominant type of nail and tack head. Rose heads are general purpose nails and were manufactured from the late 1700s until about 1820 (Azizi et al. 1996). Of course, wrought iron nails are commonly used and re-used long after the period of manufacture. Most of the architectural materials were recovered from the St. John site. This site yielded the greatest minimum numbers of items (MNI 116) and yielded also the widest range of architectural functional categories: nails, tacks, bolt, brick, tile, and window glass.

References

ALLAIRE, LOUIS. 2003. Agricultural societies in the Caribbean: The Lesser Antilles. In: Jalil Sued Badillo, ed. *General History of the Caribbean*. Volume 1, Autochthonous Societies. Paris: UNESCO Publishing; London: Macmillan. pp. 195–227.

ANDERSON-CÓRDOVA, KAREN F. 1978. *Amerindian Culture History in Trinidad: From the Late Prehistoric to Historic Periods* [typescript]. Research proposal submitted to the National Science Foundation. New Haven: Division of Anthropology, Peabody Museum of Natural History, Yale University.

ANTHONY, MICHAEL. 1988. *Towns and Villages of Trinidad and Tobago*. St. James, Port-of-Spain, Trinidad: Circle Press.

ARCHIBALD, R. DOUGLAS. 1971. *A Review of Archaeological Investigation in Trinidad* [typescript]. Port-of-Spain, Trinidad: Trinidad and Tobago Historical Society.

AZIZI, SHARLA C., DIANE DALLAL, MALLORY A. GORDON, META F. JANOWITZ, NADIA N. S. MACZAJ, AND MARIE-LORRAINE PIPES. 1996. *Analytical Coding System for Historic Period Artifacts*. [n.p.]: The Cultural Resource Group of Louis Berger and Associates.

BÉRARD, BENÔIT. 2004. *Les premières occupations agricoles de l'arc antillais, migration et insularité: Le cas de l'occupation saladoïde ancienne de la Martinique*. Oxford: Archaeopress. (British Archaeological Reports International Series 1299.)

BOOMERT, ARIE. 1984a. The Arawak Indians of Trinidad and Coastal Guiana, ca 1500–1650. *The Journal of Caribbean History* 19(2):123–188.

—1984b. *An Inventory of the Archaeological Sites in Trinidad and Tobago. Report 1: North and Central Trinidad* [mimeograph]. St. Augustine, Trinidad: University of the West Indies.

—1985. The Guayabitoid and Mayoid series: Amerindian culture history in Trinidad during late prehistoric and protohistoric times. In: Louis Allaire and Francine M. Mayer, eds. *Proceedings of the Tenth International Congress for the Study of the Pre-Columbian Cultures of the Lesser Antilles*; 1983 July 20–30; Fort-de-France, Martinique. Montreal: Centre de recherches caraïbes, Université de Montréal. pp. 93–148.

—1986. The Cayo complex of St. Vincent: Ethnohistorical and archaeological aspects of the Island Carib problem. *Antropológica* 66:3–68.

—1987. *An Inventory of the Archaeological Sites in Trinidad and Tobago. Report 2: South-East Trinidad* [mimeograph]. St. Augustine, Trinidad: University of the West Indies.

—2000. *Trinidad, Tobago and the Lower Orinoco Interaction Sphere: An Archaeological/ Ethnohistorical Study*. Alkmaar, The Netherlands: Cairi Publications.

—2001. Saladoid sociopolitical organization. In: Gérard Richard, ed. *Proceedings of the Eighteenth Congress of the International Association for Caribbean Archaeology*, Volume 1; 1999 July 11–17; True Blue, St. George, Grenada. Basse-Terre, Guadeloupe: l'Association Internationale d'Archéologie de la Caraïbe Région Guadeloupe. pp. 55–77.

—2002. Amerindian–European encounters on and around Tobago (1498–ca. 1810). *Antropológica* 97–98:71–207.

—2003. Agricultural societies in the continental Caribbean. In: Jalil Sued Badillo, ed. *General History of the Caribbean*. Volume 1, Autochthonous Societies. Paris: UNESCO Publishing; London: Macmillan. pp. 134–194.

—2004. Koriabo and the Polychrome tradition: The late-prehistoric era between the Orinoco and Amazon mouths. In: André Delpuech and Corinne L. Hofman, eds. *Late Ceramic Age Societies in the Eastern Caribbean*. Oxford: Archaeopress. pp. 251–266. (British Archaeological Reports International Series 1273.)

—2005. Golden Grove: A late-prehistoric ceramic complex of Tobago. *Leiden Journal of Pottery Studies* 21:27–60.

—2009. Between the mainland and the islands: The Amerindian cultural geography of Trinidad. *Bulletin of the Peabody Museum of Natural History* 50(1):63–73.

—2010. Crossing the Galleons' Passage: Amerindian interaction and cultural (dis)unity between Trinidad and Tobago. In: Corinne L. Hofman and Alistair J. Bright, eds. Mobility and Exchange from a Pan-Caribbean Perspective. *Journal of Caribbean Archaeology*, Special Publication 3:106–121.

BOOMERT, ARIE AND PETER O'BRIEN HARRIS. 1988. *An Inventory of the Archaeological Sites in Trinidad and Tobago* [mimeograph]. St. Augustine, Trinidad: University of the West Indies.

BORDE, PIERRE-GUSTAVE-LOUIS. 1982. *The History of the Island of Trinidad under the Spanish Government*. Volume 2, 1622–1797. Translated from the French by A. S. Mavrogordato. Newtown, Port-of-Spain, Trinidad: Paria Publishing. First published Paris, 1883.

BRADLEY, CHARLES S. 2000. Smoking pipes for the archaeologist. In: Karlis Karklins, ed. *Studies in Material Culture Research*. California, PA: Department of Anthropology, California, University of Pennsylvania and Society for Historical Archaeology. pp. 104–133.

BRERETON, BRIDGET. 1981. *A History of Modern Trinidad, 1782–1962*. London: Heinemann.

BRIGHT, ALISTAIR J. 2006. *Auger Testing at Cedros* [typescript]. Leiden: Faculty of Archaeology, Leiden University.

BULLBROOK, JOHN A. [n.d.]. *Erin Midden. Human Remains* [typescript]. Port-of-Spain, Trinidad: Royal Victoria Institute Museum.

—1927. The aborigines of Trinidad. In: Alfred Richards, ed. *Discovery Day Celebration 1927–Souvenir*. Port-of-Spain, Trinidad: Franklin's. pp. 66–72.

—1940. The Ierian race. In: Historical Society of Trinidad and Tobago. *Public Lectures Delivered Under the Auspices of the Historical Society of Trinidad and Tobago During the Session, 1938–39*. Trinidad and Tobago: A. L. Rhodes. pp. 3–46.

—1941. *The Aboriginal Remains of Trinidad and the West Indies*. Port-of-Spain, Trinidad: Trinidad and Tobago Historical Society .

—1949. The aboriginal remains of Trinidad and the West Indies. *Caribbean Quarterly* 1(1):16–21, 1(2):10–15.

—1953. *On the Excavation of a Shell Mound at Palo Seco, Trinidad, B.W.I.* New Haven: Published for the Department of Anthropology, Yale University, by Yale University Press. (Yale University Publications in Anthropology 50.)

—1960. *The Aborigines of Trinidad*. Port-of-Spain, Trinidad: Royal Victoria Institute Museum. (Occasional Paper 2.)

—1961a. *An Enquiry as to the Means for Conservation of Antiquies in the West Indies: Aboriginal Remains in Trinidad, W.I.* [mimeograph]. Port-of-Spain, Trinidad: Royal Victoria Institute Museum.

—1961b. *Premier Congrès International d'Etudes des Civilisations Pre-Colombien aux Petites Antilles, Fort-de-France, Martinique 3–8 Juillet 1961: Comments* [typescript]. Port-of-Spain, Trinidad: Royal Victoria Institute Museum.

BULLEN, ADELAIDE K. 1970. Case study of an Amerindian burial with grave goods from Grande Anse, St. Lucia. In: Ripley P. Bullen, ed. *Proceedings of the Third International Congress for the Study of the Pre-Columbian Cultures of the Lesser Antilles*; 1969 July 7–11; Grenada. St. George's: Grenada National Museum. pp. 45–60.

CARROCERA, PADRE BUENAVENTURA DE. 1968. *Misión de los Capuchinos en Cumaná*. Volume 1, Su Historia. Caracas: Academia Nacional de la Historia. (Biblioteca de la Academia Nacional de la Historia, Fuentes para la historia colonial de Venezuela 88–90.)

—1979. *Misión de los Capuchinos en Guayana*. Volume 1, Introducción y resumen histórico, documentos (1682–1758). Caracas: Academia Nacional de la Historia. (Biblioteca de la Academia Nacional de la Historia, Fuentes para la historia colonial de Venezuela 139–141.)

CARRUTHERS, CLIVE. 2003. Spanish *botijas* or olive jars from the Santo Domingo monastery, La Antiqua, Guatemala. *Historical Archaeology* 37(4):40–55.

COLLENS, JAMES HENRY. 1888. *A Guide to Trinidad: A Handbook for the Use of Tourists and Visitors*, 2nd ed. rev. and illus. London: Elliot Stock.

Concerning the Island of Trinidad Which Bears North-South in Relation to the Dragon's Mouth [typescript]. [1780]. Translated from the Spanish by Kemlin M. Lawrence. St. Augustine, Trinidad: University of the West Indies.

CRUXENT, JOSÉ M. AND IRVING ROUSE. 1958–1959. *An Archeological Chronology of Venezuela*. Washington, DC: Pan American Union. 2 vols. (Social Science Monographs 6.)

CURET, L. ANTONIO. 2004. Island archaeology and units of analysis in the study of ancient Caribbean societies. In: Scott M. Fitzpatrick, ed. *Voyages of Discovery: The Archaeology of Islands*. Westport, CT: Praeger. pp. 187–201.

—2005. *Caribbean Paleodemography: Population, Culture History, and Sociopolitical Processes in Ancient Puerto Rico*. Tuscaloosa, AL: The University of Alabama Press.

DEAGAN, KATHLEEN. 1987. *Artifacts of the Spanish Colonies of Florida and the Caribbean 1500–1800*. Volume 1, Ceramics, Glassware, and Beads. Washington, DC: Smithsonian Institution Press.

DE BOOY, THEODOOR. 1917. Certain archaeological investigations in Trinidad, British West Indies. *American Anthropologist* 19(4):471–486.

DORST, MARC C. 2004. Manzanilla 1: Creating a site-scale pottery classification at a multi-component Ceramic Age site on Trinidad. *Leiden Journal of Pottery Studies* 20:53–74.

—2006. *Preliminary Research Report: Excavation of the Amerindian SAN 1-site at Manzanilla, Trinidad February 2006*. Port-of-Spain, Trinidad: National Archaeological Committee of Trinidad and Tobago.

—2008. *The Pre-Columbian SAN 1-site, Manzanilla, Trinidad: Preliminary Research Report Fieldwork October 2007*. Port-of-Spain, Trinidad: National Archaeological Committee of Trinidad and Tobago.

DORTA, ENRIQUE MARCO, ed. 1967. *Materiales para la Historia de la Cultura en Venezuela (1523–1828). Documentos del Archivo General de Indias de Sevilla*. Caracas/Madrid: Gráficas Cóndor.

DUYMELINCK, PETRUS J. M. AND ALISTAIR J. BRIGHT. 2005. *Opgravingsverslag Cedros-Site in Trinidad van 31 Juli t/m 23 Augustus 2005* [typescript]. Leiden: Faculty of Archaeology, Leiden University.

ESPINET, ADRIAN. 1950 July 30. Archaeological finds taken to the U.S.A.: 300 persons visit South Museum since 1947. *Sunday Guardian*, Port-of-Spain, Trinidad.

FABER-MORSE, BIRGIT. 1989. Saladoid remains and adaptive strategies in St. Croix, Virgin Islands. In: Peter E. Siegel, ed. *Early Ceramic Population Lifeways and Adaptive Strategies in the Caribbean.* Oxford: British Archaeological Reports. pp. 36–45. (British Archaeological Reports International Series 506.)

—2007. The Cedros site and complex: The earliest known ceramic complex in Trinidad. In: Basil Reid, Henry Petitjean Roget and L. Antonio Curet, eds. *Proceedings of the Twenty-first Congress of the International Association for Caribbean Archaeology,* Volume 1; 2005 July 25–30; St. Augustine, Trinidad. St. Augustine, Trinidad: School of Continuing Studies, University of the West Indies. pp. 295–305.

—2010. The Palo Seco site and complex: A multi-component Ceramic Age settlement in Trinidad. In: June Heath, Jasinth Simpson and Debra-Kay Palmer, eds. *Proceedings of the Twenty-second Congress of the International Association for Caribbean Archaeology;* 2007 July 23–29; Kingston, Jamaica. Kingston: The Jamaica National Heritage Trust. pp. 196–210.

Feriz, Hans. 1959. *Zwischen Mexico und Peru.* Amsterdam: Koninklijk Instituut voor de Tropen. 2 vols. (Afdeling Culturele en Physische Anthropologie 63.)

Fewkes, Jesse Walter. 1907. The Aborigines of Porto Rico and Neighboring Islands. In: Bureau of American Ethnology. *Twenty-Fifth Annual Report of the Bureau of American Ethnology to the Secretary of the Smithsonian Institution, 1903–04.* Washington, DC: Government Printing Office. pp. 221–284.

—1914. Prehistoric objects from a shell-heap at Erin Bay, Trinidad. *American Anthropologist* 16(2):200–220.

—1922. A Prehistoric Island Culture Area of America. In: Bureau of American Ethnology. *Thirty-Fifth Annual Report of the Bureau of American Ethnology to the Secretary of the Smithsonian Institution, 1912–13.* Washington, DC: Government Printing Office. pp. 35–271.

Gassón, Rafael A. 2002. Orinoquia: The archaeology of the Orinoco River basin. *Journal of World Prehistory* 16(3):237–311.

Glazier, Stephen D. 1982. The St. Joseph and Mayo collections from Trinidad, West Indies. *The Florida Anthropologist* 35(4):208–215.

Goggin, John M. Papers. Special and Area Studies Collections, George A. Smathers Libraries, University of Florida, Gainesville, Florida.

—1960. *The Spanish Olive Jar: An Introductory Study.* New Haven: Department of Anthropology, Yale University. (Yale University Publications in Anthropology 62.)

—1968. *Spanish Majolica in the New World: Types of the Sixteenth to Eighteenth Centuries.* New Haven: Department of Anthropology, Yale University. (Yale University Publications in Anthropology 72.)

Grouard, Sandrine. 1998. *Manzanilla-Trinidad: Premier aperçu sur les Vertébrés* [typescript]. Paris: Musée national d'Histoire naturelle.

Guide to the Collections from the West Indies. 1922. New York: Museum of the American Indian, Heye Foundation. (Indian Notes and Monographs.)

Harris, Peter O'Brien. 1971. *Banwari Trace: Preliminary Report on a Pre-Ceramic Site in Trinidad, West Indies* [typescript]. Pointe-à-Pierre, Trinidad: Trinidad and Tobago Historical Society (South Section). On file at the West Indiana Section, Main Library, University of the West Indies, St. Augustine, Trinidad.

—1972. *Notes on Trinidad Archaeology* [mimeograph]. Pointe-à-Pierre, Trinidad: Trinidad and Tobago Historical Society (South Section).

—1973. Preliminary report on Banwari Trace, a preceramic site in Trinidad. In: Ripley P. Bullen, ed. *Proceedings of the Fourth International Congress for the Study of the Pre-Columbian Cultures of the Lesser Antilles;* 1971 July 26–30; Castries, St. Lucia. Castries, St. Lucia: St. Lucia Archaeological and Historical Society. pp. 115–125.

—1974. Summary of Trinidad archaeology 1973. In: Ripley P. Bullen, ed. *Proceedings of the Fifth International Congress for the Study of the Pre-Columbian Cultures of the Lesser Antilles*; 1973 July 22–28; Antigua. Antigua: The Antigua Archaeological Society. pp. 110–116.

—1976. The preceramic period in Trinidad. In: L. Sickler Robinson, ed. *Proceedings of the First Puerto Rican Symposium on Archaeology*; 1973; Santurce, Puerto Rico. San Juan: Fundacíon Arqueológica, Antropológica e Historica de Puerto Rico. pp. 33–65.

—1978. A revised chronological framework for ceramic Trinidad and Tobago. In: Jean Benoist and Francine M. Mayer, eds. *Proceedings of the Seventh International Congress for the Study of the Pre-Columbian Cultures of the Lesser Antilles*; 1977 July 11–16; Caracas, Venezuela. Montréal: Centre de recherches caraïbes, Université de Montréal. pp. 47–63.

—1985. A comparison of three Bontour period sites, Trinidad. In: Louis Allaire and Francine M. Mayer, eds. *Proceedings of the Tenth International Congress for the Study of the Pre-Columbian Cultures of the Lesser Antilles*; 1983 July 20–30; Martinique. Montreal: Centre de recherches caraïbes, Université de Montréal. pp. 265–286.

—1991a. Amerindian Trinidad and Tobago. In: Linda Sickler Robinson, ed. *Proceedings of the Twelfth Congress of the International Association for Caribbean Archaeology*; 1987 July–August; Cayenne, French Guiana. Martinique: Association Internationale d'Archéologie de la Caraïbe. pp. 259–267.

—1991b. A Paleo-Indian stemmed point from Trinidad, West Indies. In: Alissandra Cummins and Phillipa King, eds. *Proceedings of the Fourteenth Congress of the International Association for Caribbean Archaeology*; 1991; Barbados. Bridgetown: Barbados Museum and Historical Society. pp. 73–93.

HOFMAN, CORINNE L. 1993. *In Search of the Native Population of Pre-Columbian Saba (400–1450 A.D.). Part One, Pottery Styles and Their Interpretations* [dissertation]. Leiden: Leiden University.

HOFMAN, CORINNE L. AND ALISTAIR J. BRIGHT. 2004. From Suazoid to Folk pottery: Pottery manufacturing traditions in a changing social and cultural environment on St. Lucia. *New West Indian Guide/Nieuwe West-Indische Gids* 78(1–2):73–104.

HOFMAN, CORINNE L., A. J. DAAN ISENDOORN, AND MATHIJS A. BOODEN. 2005. Clays collected. Towards an identification of source areas for clays used in the production of pre-Columbian pottery in the northern Lesser Antilles. *Leiden Journal of Pottery Studies* 21:9–26.

HOFMAN, CORINNE L., A. J. DAAN ISENDOORN, MATHIJS A. BOODEN, AND LOE F. H. C. JACOBS. 2008. In tuneful threefold: Combining conventional archaeological methods, archaeometric techniques, and ethnoarchaeological research in the study of precolonial pottery of the Caribbean. In: Corinne L. Hofman, Menno L. P. Hoogland and Annelou L. van Gijn, eds. *Crossing the Borders: New Methods and Techniques in the Study of Archaeological Materials from the Caribbean*. Tuscaloosa: University of Alabama Press. pp. 21–33.

HOFMAN, CORINNE L. AND LOE F. H. C. JACOBS. 2004. Different or alike? A technological comparison between late-prehistoric ceramics and modern-day Folk pottery of St. Lucia (W.I.). *Leiden Journal of Pottery Studies* 20:7–22.

HOWARD, GEORGE D. 1943. *Excavations at Ronquín, Venezuela*. New Haven: Published for the Department of Anthropology, Yale University, by Yale University Press. (Yale University Publications in Anthropology 28.)

IRWIN, GEOFFREY. 1992. *The Prehistoric Exploration and Colonisation of the Pacific*. Cambridge: Cambridge University Press.

JONES, OLIVE R. 1986. *Cylindrical English Wine and Beer Bottles, 1735–1850*. Ottawa, ON: National Historic Parks and Sites Branch, Environment Canada Parks Service.

JONES, OLIVE R. AND CATHERINE SULLIVAN. 1989. *The Parks Canada Glass Glossary for the description of containers, tableware, flat glass, and closures,* 2nd ed. Ottawa, ON: National Historic Parks and Sites Branch, Environment Canada Parks Service.

JOSEPH, EDWARD L. 1838. *History of Trinidad.* Port-of-Spain, Trinidad: H.J.Mills.

KEEGAN, WILLIAM F. 1995. Modeling dispersal in the prehistoric West Indies. *World Archaeology* 26(3):400–420.

—2001. Archaeological investigations on Ile à Rat, Haiti: Avoid the oid. In: Gérard Richard, ed. *Proceedings of the Eighteenth Congress of the International Association for Caribbean Archaeology,* Volume 2; 1999 July 11–17; True Blue, St. George, Grenada. Basse-Terre, Guadeloupe: l'Association Internationale d'Archéologie de la Caraïbe Région Guadeloupe. pp. 233–239.

—2004. Islands of chaos. In: André Delpuech and Corinne L. Hofman, eds. *Late Ceramic Age Societies in the Eastern Caribbean.* Oxford: Archaeopress. pp. 33–44. (British Archaeological Reports International Series 1273.)

KEEGAN, WILLIAM F. AND BRYAN BYRNE. 2001. Structural analysis of Saladoid adornos from Grenada. In: Gérard Richard, ed. *Proceedings of the Eighteenth Congress of the International Association for Caribbean Archaeology,* Volume 1; 1999 July 11–17; True Blue, St. George, Grenada. Basse-Terre, Guadeloupe: l'Association Internationale d'Archéologie de la Caraïbe Région Guadeloupe. pp. 21–23.

KINGSLEY, CHARLES. 1889. *At Last: A Christmas in the West Indies,* 3rd ed. London: Macmillan.

KYBERG, LYZ. 1976. *Polychrome Painting at Cedros?* [typescript]. New Haven: Division of Anthropology, Peabody Museum of Natural History, Yale University.

LANGE, AMANDA E. 2001. *Delftware at Historic Deerfield, 1600–1800.* Deerfield, MA: Historic Deerfield.

LIAÑO, LICENCIADO PEDRO DE. 1596 March 15. Despatch from Licenciado Pedro de Liano to the King of Spain [reprint]. Port-of-Spain: The Trinidad Historical Society. (Publication 23.) Available at: http://ufdc.ufl.edu/UF00080962/00011

LISTER, FLORENCE C. AND ROBERT H. LISTER. 1974. Maiolica in Colonial Spanish America. *Historical Archaeology* 8:17–52.

—1976. *A Descriptive Dictionary for 500 Years of Spanish-Tradition Ceramics [13th Through 18th Centuries].* Society for Historical Archaeology. (Special Publication 1.)

—1982. *Sixteenth Century Maiolica Pottery in the Valley of Mexico.* Tucson, AZ: University of Arizona Press. (Anthropological Papers of the University of Arizona 39.)

MAHARAJ, DAVID D. A. 1996. *Excavation Report Mayo (Ceramic Site, Trinidad)* [manuscript]. San Fernando, Trinidad: Trinidad and Tobago Historical Society (South Section).

MALLET, F. 1802. *Descriptive Account of the Island of Trinidad.* London: W. Faden.

MALLIOS, SETH. 2005. Back to the bowl: Using English tobacco pipebowls to calculate mean site-occupation dates. *Historical Archaeology* 39(2):89–104.

MARKEN, MITCHELL W. 1994. *Pottery from Spanish Shipwrecks, 1500–1800.* Gainesville, FL: University Press of Florida.

McCASHION, JOHN H. 1979. A preliminary chronology and discussion of seventeenth and early eighteenth century clay tobacco pipes from New York State sites. In: Peter Davey, ed. *The Archaeology of the Clay Tobacco Pipe.* Oxford: British Archaeological Reports. pp. 63–150. (British Archaeological Reports International Series 60.)

MILLER, GEORGE L. 1989. *A Chronology of English Shell Edged Pearl and White Wares.* Williamsburg, VA: Published by the author.

—2000. A revised set of CC Index Values for classification and economic scaling of English ceramics from 1787 to 1880. In: David R. Brauner, ed. *Approaches to Material Culture: Research for Historical Archaeologists*, 2nd ed. California, PA: Society for Historical Archaeology. pp. 86–110.

MILLER, GEORGE L., PATRICIA SAMFORD, ELLEN SHLASKO, AND ANDREW MADSEN. 2000. Telling time for archaeologists. *Northeast Historical Archaeology* 29:1–22.

MILLER II, J. JEFFERSON. 1974. *English Yellow-Glazed Earthenware*. Washington, DC: Smithsonian Institution Press.

MORAVETZ, IOSIF. 2005. *Imaging Adornos: Classification and Iconography of Saladoid Adornos from St. Vincent, West Indies*. Oxford: Archaeopress. (British Archaeological Reports International Series 1445.)

NEWSON, LINDA A. 1976. *Aboriginal and Spanish Colonial Trinidad: A Study in Culture Contact*. London: Academic Press.

NIEWEG, DENNIS C. AND MARC C. DORST. 2001. The Manzanilla 1 (SAN-1) site, Trinidad. In: Luc Alofs and Raymundo A. C. F. Dijkhoff, eds. *Proceedings of the Nineteenth Congress of the International Association for Caribbean Archaeology*, Volume 1; 2001 July 22–28; Aruba. Aruba: Museo Arqueologico Aruba. pp. 173–185. (Publication of the Archaeological Museum Aruba 9.)

NOËL HUME, IVOR. 1985. *A Guide to Artifacts of Colonial America*. New York: Knopf.

Notarial Record of the Founding of the Town of San Josephe de Oruna. 1592 October 3 [reprint]. Port-of-Spain: The Trinidad Historical Society. (Publication 15.) Available at: http://ufdc.ufl.edu/UF00080962/00003

OLSEN, FRED. 1969a. *Excavations in Trinidad: Sept. 19 1969: CEDROS* [typescript]. New Haven: Division of Anthropology, Peabody Museum of Natural History, Yale University.

—1969b. *Excavations at Palo Seco Site, Trinidad: Sept 17, 1969* [typescript]. New Haven: Division of Anthropology, Peabody Museum of Natural History, Yale University.

—1973. On the trail of the Arawaks: When did they arrive in Trinidad? In: Ripley P. Bullen, ed. *Proceedings of the Fourth International Congress for the Study of the Pre-Columbian Cultures of the Lesser Antilles*; 1971 July 26–30; Castries, St. Lucia. Castries, St. Lucia: St. Lucia Archaeological and Historical Society. pp. 181–191.

—1974. *On the Trail of the Arawaks*. Norman, OK: University of Oklahoma Press.

OSGOOD, CORNELIUS C. 1942. Prehistoric contact between South America and the West Indies. *Proceedings of the National Academy of Sciences* 28(1):1–4.

OSWALD, ADRIAN. 1975. *Clay Pipes for the Archaeologist*. Oxford: British Archaeological Reports. (British Archaeological Reports International Series 15.)

OUSIEL, JACQUES. 1637 December. A Report by Ousiel, Secretary of Tobago to the Dutch West India Company [reprint]. Port-of-Spain: The Trinidad Historical Society. (Publication 137.) Available at: http://ufdc.ufl.edu/UF00080962/00125

PETERSEN, JAMES B. 2004. *Trinidad Ceramics* (Notes by Jim Petersen) Peabody Museum, Yale Univ., July 21–2 [manuscript]. New Haven: Division of Anthropology, Peabody Museum of Natural History, Yale University.

PHILIP, JOHN BAPTISTA. 1824. *An Address to the Right Honourable Earl Bathurst Relative to the Claims Which the Coloured Population of Trinidad Have to the Same Civil and Political Privileges with their White Fellow Subjects, by a Free Mulatto of the Island*. London: Printed for J. Hatchard.

PINCHON, P. ROBERT. 1952. Introduction à l'archéologie Martiniquaise. *Journal de la Société des Américanistes* 41:305–352.

PRESTON, RICHARD S., ELAINE PERSON, AND EDWARD. S. DEEVEY. 1955. Yale Natural Radiocarbon Measurements II. *Science* 122:954–960.

REID, BASIL A. 2004. Reconstructing the Saladoid religion of Trinidad and Tobago. *The Journal of Caribbean History* 38(2):243–278.

RICKARD, JONATHAN. 2006. *Mocha and Related Dipped Wares, 1770–1939.* Hanover, NH: University Press of New England.

RICE, PRUDENCE M. 1987. *Pottery Analysis: A Sourcebook.* Chicago: University of Chicago Press.

ROUSE, IRVING. Papers. Division of Anthropology Archives, Peabody Museum of Natural History, Yale University.

—1947. Prehistory of Trinidad in relation to adjacent areas. *Man* 47:93–98.

—1948. The Arawak. In: Julian H. Steward, ed. *Handbook of South American Indians.* Volume 4, The Circum-Caribbean Tribes, Part 3, The West Indies. Washington, DC: Smithsonian Institution. pp. 507–546. (Bureau of American Ethnology Bulletin 143.)

—1951. Prehistoric Caribbean culture contact as seen from Venezuela. *Transactions of the New York Academy of Sciences* 2(13):342–347.

—1953. Appendix B: Indian Sites in Trinidad. In: John A. Bullbrook. *On the Excavation of a Shell Mound at Palo Seco, Trinidad, B.W.I.* New Haven: Published for the Department of Anthropology, Yale University, by Yale University Press. (Yale University Publications in Anthropology 50.) pp. 99–111.

—1960. *The Entry of Man into the West Indies.* New Haven: Department of Anthropology, Yale University. (Yale University Publications in Anthropology 61.)

—1964. Prehistory of the West Indies. *Science* 144:499–514.

—1978. The La Gruta sequence and its implications. In: Erika Wagner and Alberta Zucchi, eds. *Unidad y Variedad: Ensayos Antropológicos en Homenaje a José M. Cruxent.* Caracas: Ediciones del Centro de Estudios Avanzados, Instituto Venezolano de Investigaciones Científicas. pp. 203–229

—1983. Diffusion and interaction in the Orinoco Valley and on the Coast. In: Louis Allaire and Francine M. Mayer, eds. *Proceedings of the Ninth International Congress for the Study of the Pre-Columbian Cultures of the Lesser Antilles*; 1981 August 2–8; Santo Domingo, Dominican Republic. Montreal: Centre de recherches caraïbes, Université de Montréal. pp. 3–13.

—1986. *Migrations in Prehistory: Inferring Population Movement from Cultural Remains.* New Haven: Yale University Press.

—1992. *The Tainos: Rise and Decline of the People Who Greeted Columbus.* New Haven: Yale University Press.

ROUSE, IRVING AND LOUIS ALLAIRE. 1978. Caribbean. In: R. E. Taylor and Clement W. Meighan, eds. *Chronologies in New World Archaeology.* New York: Academic Press. pp. 431–481.

ROUSE, IRVING AND BIRGIT FABER-MORSE. 1999. *Excavations at the Indian Creek Site, Antigua, West Indies.* New Haven: Department of Anthropology and Peabody Museum of Natural History, Yale University. (Yale University Publications in Anthropology 82.)

ROUSE, IRVING, BIRGIT FABER-MORSE, AND DESMOND V. NICHOLSON. 1995. Excavations at Freeman's Bay, Antigua. In: Ricardo E. Alegría and Miguel Rodríguez, eds. *Proceedings of the Fifteenth Congress of the International Association for Caribbean Archaeology*; 1993 July 25–31; San Juan, Puerto Rico. San Juan: Centro de Estudios Avanzados de Puerto Rico y el Caribe/ Fundación Puertorriqueña de las Humanidades/Universidad del Turabo. pp. 445–457.

RYE, OWEN S. 1981. *Pottery Technology: Principles and Reconstruction.* Washington, DC: Taraxacum.

SALAZAR, PEDRO DE. 1595 July 10. Despatch from Pedro de Salazar to the King of Spain [reprint]. Port-of-Spain: The Trinidad Historical Society. (Publication 21.) Available at: http://ufdc.ufl.edu/UF00080962/00009

SAMFORD, PATRICIA M. 2000. Response to a market: Dating English Underglazed Transfer-printed wares. In: David R. Brauner, ed. *Approaches to Material Culture Research for Historical Archaeologists,* 2nd ed. California, PA: Society for Historical Archaeology. pp. 56–85.

SANOJA OBEDIENTE, MARIO AND IRAIDA VARGAS ARENAS. 1983. New light on the prehistory of eastern Venezuela. In: Fred Wendorf and Angela E. Close, eds. *Advances in World Archaeology,* Volume 2. New York: Academic Press. pp. 205–244.

SAUNDERS, NICHOLAS J. AND ARCHIBALD S. CHAUHARJASINGH. 2003. *An Inventory of Archaeological Sites in Trinidad and Tobago. Report 4: Counties of Victoria and St. Patrick* [typescript]. London: University College London.

SERRAND, NATHALIE. 2001. Occurrence of exogenous freshwater bivalves (Unionoida) in the Lesser Antilles during the 1st millennium A.D.: Example from the Hope Estate Saladoid site (St. Martin, French Lesser Antilles). In: Gérard Richard, ed. *Proceedings of the Eighteenth Congress of the International Association for Caribbean Archaeology,* Volume 1; 1999 July 11–17; True Blue, St. George, Grenada. Guadeloupe, French West Indies: pp. 136–152.

SHEPARD, ANNA O. 1968. *Ceramics for the Archaeologist,* 6th ed. Washington, DC: Carnegie Institution. (Publication 609.)

SILVER, ANNETTE. 2009. The English ceramic assemblage at the St. Joseph site, Trinidad. *Bulletin of the Peabody Museum of Natural History* 50(1):199–208.

SILVER, ANNETTE AND BIRGIT FABER-MORSE. 2010. The Spanish ceramic assemblage at the St. Joseph site, Trinidad. *Proceedings of the Twenty-second Congress of the International Association for Caribbean Archaeology;* 2007 July 23–29; Kingston, Jamaica. Kingston: The Jamaica National Heritage Trust. pp. 495–507.

SLEIGHT, FREDERICK W. 1946. Notes on a find from Trinidad. *American Antiquity* 11(4):260–261.

STUIVER, MINZE, PAULA J. REIMER, AND RON W. REIMER. 2005. CALIB 14C Calibration Program. CALIB 5.0. [online program and documentation]. Available from: http://calib.qub.ac.uk/calib/

SUSSMAN, LYNNE. 1997. *Mocha, Banded, Cat's Eye, and Other Factory-Made Slipware.* Boston: Council for Northeast Historical Archaeology. (Studies in Northeast Historical Archaeology 1.)

—2000. British military tableware, 1760–1830. In: David R. Brauner, ed. *Approaches to Material Culture Research for Historical Archaeologists,* 2nd ed. California, PA: Society for Historical Archaeology. pp. 44–55.

TELLES, GEORGE A. 1927. The ancient capital of Trinidad: St. Joseph—A town of many memories. In: Alfred Richards, ed. *Discovery Day Celebration 1927—Souvenir.* Port-of-Spain, Trinidad: Franklin's. pp. 60–62.

VARGAS ARENAS, IRAIDA. 1979. *La Tradición Saladoide del Oriente de Venezuela: La Fase Cuartel.* Caracas: Academia Nacional de la Historia. (Biblioteca de la Academia Nacional de la Historia, Serie Estudios, Monografías y Ensayos 5.)

[TTHS] TRINIDAD AND TOBAGO HISTORICAL SOCIETY (SOUTH SECTION). *Annual Reports; Newsletters.* Port-of-Spain, Trinidad and Tobago, West Indies: Government Printing Office.

VERTEUIL, LOUIS A. A. GASTON DE. 1884. *Trinidad: Its Geography, Natural Resources, Administration, Present Condition, and Prospects*, 2nd ed. London: Cassell.

WALL, GEORGES P. AND JAMES G. SAWKINS. 1860. *Report on the Geology of Trinidad; or, Part I of the West Indian Survey*. London: Printed for H.M. Stationery Off.: Longman, Green, Longman, and Roberts, (Memoirs Geological Survey.)

WILBERT, JOHANNES. 1993. *Mystic Endowment: Religious Ethnography of the Warao Indians*. Cambridge: Harvard University Press.

WILKIE, LAURIE A. 2000. Glass-knapping at a Louisiana plantation: African-American tools? In: David R. Brauner, ed. *Approaches to Material Culture Research for Historical Archaeologists*, 2nd ed. California, PA: Society for Historical Archaeology. pp. 189– 201.

WING, ELIZABETH S. 1962. *Succession of Mammalian Faunas on Trinidad, West Indies* [dissertation]. Gainesville, FL: University of Florida. Available from: ProQuest Dissertations and Theses [database online]; http://www.proquest.com (publication no. 6700376).

—1977. Factors influencing exploitation of marine resources. In: Elizabeth P. Benson, ed. *The Sea in Pre-Columbian World;* a conference at Dumbarton Oaks, October 26th and 27th, 1974. Washington, DC: Dumbarton Oaks Research Library and Collections. pp. 47– 64.

WING, ELIZABETH S. AND ELIZABETH J. REITZ. 1982. Prehistoric fishing economies of the Caribbean. *Journal of New World Archaeology* 5(2):13–32.

WISE, KENRICK S. 1934–1938. *Historical Sketches of Trinidad and Tobago*. London: Printed for the Historical Society of Trinidad and Tobago by Baines and Scarsbrook, Ltd. 4 vols.